MATERIALS SCIENCE AND TECHNOLOGIES

CERAMIC MATERIALS

SYNTHESIS, PERFORMANCE AND APPLICATIONS

Materials Science and Technologies

Additional books in this series can be found on Nova's website under the Series tab.

Additional E-books in this series can be found on Nova's website under the E-book tab.

MATERIALS SCIENCE AND TECHNOLOGIES

CERAMIC MATERIALS

SYNTHESIS, PERFORMANCE AND APPLICATIONS

JACQUELINE PEREZ
EDITOR

New York

NOTICE TO THE READER

Library of Congress Cataloging-in-Publication Data

Names: Perez, Jacqueline, editor.
Title: Ceramic materials: synthesis, performance and applications / (editor) Jacqueline Perez.
Description: Hauppauge, New York: Nova Science Publisher's, Inc., 2016. | Series: Materials science and technologies | Includes index.
Identifiers: LCCN 2016035623 (print) | LCCN 2016038286 (ebook) | ISBN 9781634859653 (hardcover) | ISBN 9781634859806
Subjects: LCSH: Ceramic materials.
Classification: LCC TA430 .C46 2016 (print) | LCC TA430 (ebook) | DDC 620.1/4--dc23
LC record available at https://lccn.loc.gov/2016035623

Published by Nova Science Publishers, Inc. † New York

CONTENTS

PREFACE

This book discusses the synthesis, performance and applications of ceramic materials. Chapter One presents the recycling of biomass ashes in the obtaining of clay bricks for possible use as construction material. Chapter Two summarizes the use of nanostructured ceramics for the control of heat flows. Chapter Three deals with analytical modelling of thermal stresses in a multi-particle-matrix system with isotropic cylindrical particles which are periodically distributed in an isotropic infinite matrix. Chapter Four deals with analytical modelling of thermal stresses which originate during a cooling process of an elastic solid continuum. Chapter Five provides recent information on the use of zirconia in dentistry, its characteristics and indications, with a particular emphasis on surface conditioning methods to promote adhesion of resin-based materials to zirconia.

In Chapter 1, various fly and bottom biomass ash, such as: rice husk fly ash, wood bottom ash, fly and bottom pine-olive pruning ash, olive stone bottom ash, olive pomace bottom and fly ash, were individually blended with clay to produce ceramic bricks. The clay and biomass ash waste were characterized by X-ray fluorescence, elemental chemical analysis and thermogravimetric and differential thermal analysis. The bricks were manufactured by mixing clay and 20 wt% of ash. These bricks were fired at 900ºC. The influence of the type of biomass ash added on the linear shrinkage, bulk density, water absorption, apparent porosity and mechanical properties was investigated. The results indicated that the addition of biomass ash decreased the bulk density and the compressive strength (25% lower) although the incorporation of fly rice husk ash produced a maximum decreased of 60.8%. Instead it was observed an increment in water absorption from 4.0 to 17.3% with respect to the control bricks containing only clay. Heavy metal concentration in the leachates of the biomass ash-clay bricks was far lower than those recommended in the Spanish legislation (OM AAA/661/2013) and US-EPA standard. Therefore, fired bricks fulfill standard requirements for clay masonry units, offering, at the same time, good mechanical properties.

The recycling of these biomass ashes in the obtaining of clay bricks for possible use as construction material is presented as a potentially feasible solution according to the technical qualities of the manufactured bricks, as well as, the economic and environmental benefits that implies its use in replacement of natural raw materials.

As described in Chapter 2, nowadays considerable attention is paid to the creation of a new type of nanostructured materials in which one can control the heat flow. Since it is believed that the basic element of such thermocrystals should be phononic lattices with a wide forbidden gap, it is an urgent task to find new methods for their synthesis that strikes a balance between the thermocrystal efficiency and ease of fabrication.

It is shown both theoretically and experimentally that the compacted ceramics can exhibit the properties of a phononic lattice, i.e., a forbidden gap may arise in the phonon spectrum. The position and width of the gap in such systems are determined by the average grain size of ceramics, as well as by the thickness and elastic properties of the grain boundaries. Inclusions of metal phase in dielectric matrix can create photonic traps for the nanocomposite materials and determine their diffusion ratio.

The approach for the synthesis of nanocomposites by use of ceramics technology was introduced. It allows to create materials with required characteristics by the compaction and sintering process and to give a practical advice on synthesis of nanostructured ceramics with specified parameters.

Chapter 3 deals with analytical modelling of thermal stresses in a multi-particle-matrix system with isotropic cylindrical particles which are periodically distributed in an isotropic infinite matrix. This multi-particle-matrix system represents a model system which is applicable to real two-component materials of with precipitates of a cylindrical shape. The thermal stresses as functions of microstructural parameters (particle volume fraction, particle radius, inter-particle distance) originate during a cooling process as a consequence of the difference in thermal expansion coefficients of the cylindrical particle and the matrix. The analytical modeling is based on an application of suitable mathematical techniques on fundamental equations of mechanics of solid elastic continuum. The mathematical techniques thus result in analytical solutions for both the cylindrical particle and the infinite matrix.

Finally, numerical values of the thermal stresses in a real two-component material with cylindrical precipitates are determined.

Chapter 4 deals with analytical modelling of thermal stresses which originate during a cooling process of an elastic solid continuum. This continuum consists of an isotropic infinite matrix with isotropic spherical particles and spherical pores. The particles and pores are both periodically distributed in the infinite matrix which is imaginarily divided into identical cubic cells. Each cell contains either a central particle or a central pore. This porous multi-particle-matrix system represents a model system which is applicable to porous two-component materials of a precipitate-matrix type characterized by microstructural parameters, i.e. the cubic cell dimension; radii and volume fractions of both the particles and of the pores. The thermal stresses are a consequence of different thermal expansion coefficients of the isotropic matrix and isotropic particle. Resulting from fundamental equations of mechanics of an elastic solid continuum, the thermal stresses are determined within this cell, and thus represent functions of these microstructural parameters. Finally, numerical values for a real porous two-component material of the precipitate-matrix type are obtained.

As explained in Chapter 5, Yttria-stabilized tetragonal zirconia polycrystal (Y-TZP) has been used in dentistry in order to manufacture prosthetic frameworks, monolithic crowns and implant abutments due to its superior mechanical properties, biocompatibility, chemical stability and appropriate aesthetics as opposed to other materials. The survival of dental ceramic restorations depends on durable bond strength between the restorative material, composite resin luting cement and the tooth surface. However, it is difficult to establish a durable mechanical or chemical adhesion in zirconia-based prostheses since yttrium-stabilized zirconia is an oxide ceramic that does not contain silicon dioxide (SiO_2) phase in its microstructure. In order to achieve strong and reliable adhesion between resin composite luting cements and zirconia surfaces, it is crucial to employ a method that does not impair the mechanical properties and at the same time rendering it compatible with the luting cement. Furthermore, the chosen method should be practical, easy to perform and should not cause $t \rightarrow m$ phase transformation. Several methods and protocols for conditioning zirconia surfaces prior to adhesive cementation have been suggested in the literature such as physical, physicochemical and chemical methods. Typically, while physical surface conditioning methods are based on employing air-borne particle abrasion with alumina particles, physicochemical methods use silica-coated alumina particles followed by silanization. It is also possible to activate the zirconia surface chemically using functional-monomer containing adhesive promoters in the form of adhesive cements or primers. Generally, combination of micromechanical and chemical surface conditioning methods are preferred to enhance adhesion to zirconia. As this ceramic demonstrates superior properties compared to other ceramics, it is essential to study the peculiar characteristics of dental zirconia after surface conditioning methods and suggest one that does not damage its favorable mechanical properties. Chapter 5 will provide the recent information on the use of zirconia in dentistry, its characteristics and indications, with a particular emphasis on surface conditioning methods to promote adhesion of resin-based materials to zirconia.

In: Ceramic Materials
Editor: Jacqueline Perez

ISBN: 978-1-63485-965-3

Chapter 1

EVALUATION OF FLY AND BOTTOM ASH OF DIFFERENT BIOMASS COMBUSTION AS RAW MATERIALS IN CLAY-BASED CERAMICS

D. Eliche-Quesada[1,*], M.A. Felipe-Sesé[1] and A. Infantes-Molina[2]

[1]Department of Chemical, Environmental, and Materials Engineering, Higher Polytechnic School of Jaén, University of Jaén, Jaén, Spain

[2]Department of Inorganic Chemistry, Crystallography and Mineralogy (Associate Unit to the ICP-CSIC0, Faculty of Sciences, University of Malaga, Málaga, Spain

Abstract

In this work, various fly and bottom biomass ash, such as: rice husk fly ash, wood bottom ash, fly and bottom pine-olive pruning ash, olive stone bottom ash, olive pomace bottom and fly ash, were individually blended with clay to produce ceramic bricks. The clay and biomass ash waste were characterized by X-ray fluorescence, elemental chemical analysis and thermogravimetric and differential thermal analysis. The bricks were manufactured by mixing clay and 20 wt% of

---

* Address: Department of Chemical, Environmental, and Materials Engineering, Higher Polytechnic School of Jaén, University of Jaén, Campus Las Lagunillas s/n, 23071 Jaén, Spain (Corresponding author).

ash. These bricks were fired at 900ºC. The influence of the type of biomass ash added on the linear shrinkage, bulk density, water absorption, apparent porosity and mechanical properties was investigated. The results indicated that the addition of biomass ash decreased the bulk density and the compressive strength (25% lower) although the incorporation of fly rice husk ash produced a maximum decreased of 60.8%. Instead it was observed an increment in water absorption from 4.0 to17.3% with respect to the control bricks containing only clay. Heavy metal concentration in the leachates of the biomass ash-clay bricks was far lower than those recommended in the Spanish legislation (OM AAA/661/2013) and US-EPA standard. Therefore, fired bricks fulfill standard requirements for clay masonry units, offering, at the same time, good mechanical properties.

The recycling of these biomass ashes in the obtaining of clay bricks for possible use as construction material is presented as a potentially feasible solution according to the technical qualities of the manufactured bricks, as well as, the economic and environmental benefits that implies its use in replacement of natural raw materials.

Keywords: fly and bottom biomass ash, recycling, ceramics, technological properties, sustainability

1. Introduction

Economic and social development requires energy, in a manner that the need for energy consumption in developed countries is increasing at at a rate of nearly 1% a year, whereas the rate is higher in emerging countries, approximately 5% a year [1].

It is estimated that oil and gas reserves will only cover energy needs for the next 40 and 60 years, respectively. Furthermore, emissions generated as a result of fossil fuels combustion are responsible of serious environmental problems such as global warming. As a result, renewable energy seems to be one the most effective solution due to its renewable and environmentally sustainable nature [2] in order to ensure a high quality of life and the well-being of coming generations [3].

The energy generated from biomass is one of the most promising ways of reducing, in significant quantities, CO_2 from the combustion of coal and natural gas. According to definition in Directive 2003/54/EC, "biomass" is the biodegradable fraction of products, waste and residues from biological origin from agriculture (including vegetal and animal substances), forestry and related industries including fisheries and aquaculture, as well as the biodegradable fraction of industrial and municipal waste (Directive 2003/54/EC) [4].

Spain, following the ambitious European Union objectives has assumed the challenge to establish an ambitious energy model with the intention of promoting renewable resources like biomass to fulfill the energy targets. This is the aim of the "Plan de Acción Nacional de Energías Renovables" (PANER) (2010-2020) [5].

Andalusia (Spain) is the second-largest autonomous community in Spain, with wide forest and farmland surface and is the region that records the highest consumption in the country of the most promising source of renewable energy, biomass [6], with 18 generators of electricity from residual biomass which produce 257,48 MW [7].

Solid biomass is the best known type of biomass and can be included within this group: wood obtained from silviculture and forestry activities; wastes from industries dealing with any kind of solid biomass (e.g., carpentry, paper industry…); residues obtained from the pruning and cleaning of parks and gardens; energy crops; peat; agriculture and agro-industrial waste (such as pomace, sawdust, olive stones…); and organic fraction of solid urban wastes. Different sources of solid biomass are used in Andalusia to produce electric energy: olive pomace (218,3 Ktoe), energetic crops (190,4 Ktoe), forestry residues (106,7 Ktoe) and industrial wastes (175,6 Ktoe) depicted the main sources of solid biomass in 2013 of the total quantity (723,7 ktoe) [7].

The wastes generated from the olive sector in Andalusia, is the most important source of residual biomass for electric and thermal energy. Approximately 50% of biomass is olive tree residues; the other half is principally obtained from sunflowers, cotton and fruit trees [8]. It should be born in mind that Spain is the main world olive producer (4,577,800 tonnes) [9] and the production is mainly concentrated in Andalusia. The potential exploitation of solid olive wastes for energy purposes have been extensively researched [10-15].

However, an environmental and economic problem inherent to solid biomass combustion, stems from the large quantities of fly and bottom ashes generated from the process. Ash is usually accumulated in landfills which lead to a clear damage to the land and the surrounding area, contributing to air pollution and water contamination. In addition, the occupation of a territory stops it from being productive and the disposal in landfills is also problematic due to space limitations. The composition of the ash depends on the minerals absorbed or incorporated into the biomass during cultivation and harvesting. Likewise, ash is usually composed by organic material unburned during the usual inefficient processes [16]. Accordingly, the potential reuse of the ash is determined by its physical, chemical and environmental properties. Obviously, quality and quantity of the ash are directly influenced by the characteristics of the biomass and the combustion technology employed [17]. Regarding the type of ash, bottom ash is the portion of residue that remains in the incinerator or furnace, while fly ash is the portion of ash that goes through the chimney and is retained to prevent it from being released into the atmosphere [18].

The management of the ashes produced in great quantity as a result of the combustion processes is a challenge. There have been an increasing number of studies in all over the world regarding the characterization and the establishing of appropriated processes in which ashes can be effectively reused. As a result of this research, different areas of application of ash were proposed, depending on its composition and properties. One of the most valuable components of the ash is amorphous silica, with a wide range of applications, such as manufacturing of silica gels, silicon chip, synthesis of activated carbon and silica, production of construction materials and insulation, zeolites, catalysts, ingredients for batteries, graphene, carbon capture, drug delivery vehicles [19]. In addition, biomass ashes have been studied focusing on applications as a construction material replacing, in part or completely, traditional construction materials. This has been fostered by the growing environmental awareness in the building industry due to the large quantities of raw materials needed for ceramic production [20, 21]. As a result, the incorporation of waste in place of traditional construction material, while requirements are fulfilled, is environmentally and economically attractive. Some authors have studied building materials using as raw materials ashes from different sources with satisfactory results. So, Cabrera et al. [16] studied the properties of biomass bottom ash from the combustion of wood and olive trees residues and determined that this bottom ashes had acceptable properties to be used as a filler material in the core of road embankments over 5 m in height without additional precautionary measures, such as the construction of road shoulders. Rauta et al. [22] studied brick samples prepared from paper pulp, rice husk ash and cement. The optimal composition with paper pulp (80 wt%), rice husk ash (10 wt%) and cement (10 wt%) had a higher strength and physic-chemical characteristics than conventional burnt clay bricks. Zang et al. [23] studied the incorporation of municipal solid waste fly ash in clay bricks. The optimal mixture ratio of materials was, municipal fly ash: 20 wt%, red ceramic clay: 60 wt%, feldspar: 10 wt%, gang sand: 10 wt%. And the optimal sintering temperature was 950ºC. Other authors have studied the using of ash from coal power plants [24-29]. Eliche-Quesada and Leite-Costa [30] studied the using of bottom ash from olive pomace combustion to replace different amounts (10-50 wt%) of clay in brick manufacturing. The optimal amount of bottom ashes was 10 and 20 wt% of waste to produced bricks with lower bulk density and suitable compressive strength. Fired bricks fulfil standards requirements for clay masonry units, offering, at the same time, a reduction in thermal conductivity compared to control bricks (only clay). Cheeseman et al. [31] studied the use of sewage sludge incinerator ash to form new ceramic materials. Syed et al. [32] studied the incorporation of rice husk ash or sugarcane bagasse. Clay bricks incorporating with biomass ash exhibited lower compressive strength compared to that of clay bricks without waste. However, compressive strength of bricks with 5% of rice husk ash or sugarcane bagasse satisfied the standards.

In the present study, clay bricks by means of conventional methods, using different types of biomass ashes have been manufactured. The samples of ashes were obtained from Andalusian biomass power plants that belong to the following companies: Tradema (Linares, Jaén), Herba Ricemills (San Juan de Aznalfarache, Sevilla), Aldebarán Energía del Gualdalquivir (Andújar, Jaén), Valoriza Energía and Energía La Loma S.A. (Villanueva del Arzobispo, Jaén).

Tradema is a wooden board factory located in the town of Linares (Andalusia). The residues generated during the process constituted primarily of wood scraps from the bark of conifers (52%), wood dust and non-conforming products (48%) are used as fuel in the cogeneration plant in order to obtain electrical and thermal energy with 2 MW of power [7]. Herba Ricemills, is a factory located in San Juan de Aznalfarache (Andalusia) which produces energy from a power plant by the combustion of the rice husk, the external layer of the paddy grain that is separated from the rice grains during the milling process. In Andalusia 58,7 Ktn of rice husk were obtained derived from the industrial rice production [7], becoming an interesting source of biomass available in Andalusia, especially since one ton of rice husk can generate 800 Kw h [19]. Aldebarán Energía del Gualdalquivir uses biomass as a source of energy constituted of cereal straw and wastes from the maintenance and improvement treatments of the forests and olive groves. The cogeneration plant produces 6 MW of power. The forest waste depicted 1,34 Mt in Andalusia, while residues agricultural residues from olives groves, represented 2,52 Mt in 2013 [7]. Valoriza Energía Company is located in the Scientific and Technological Park Geolit at Mengibar (Jaen, Spain) provided air conditioning to the Park using as fuel olive stone.

Olive pomace is a by-product resulting from the extraction of olive oil obtained directly from olives. From every ton of olive processed, 0,27 ton of olive oil and 0,73 ton of olive pomace will be produced. In turn, from olive pomace, oil is also recovered by a chemical process. The residue resulting from this last process is itself a potential source of renewable energy, with 4,200 kcal/kg as a calorific value. Furthermore, it should be taken into account that in a normal season, 840 Kt/year of this waste is obtained in Andalusia. On this basis, Energía La Loma S.A. uses this waste to generate 16 MW of electrical energy in its power plant [7].

In view of the above, the present work aims at providing further knowledge in the technological properties of clay bricks and biomass combustion bottom or fly ash (BA and FA) from the power plants previously mentioned. For this purpose, raw materials were characterized and clay-ash mixtures with 20 wt% of BA or FA were conformed. Green samples were fired at 900°C and the effect of type of ash was studied by physical, mechanicall and environmental properties.

2. Experimental

2.1. Preparation of the Samples

Clay was supplied by a clay pit located in Bailen, Jaen (Spain). It was obtained by mixing three types of raw clay in the following percentages: 30 wt% red, 30 wt% yellow and 40 wt% black clay. To obtain uniform particle size, the clay was crushed and ground to yield a powder with a particle size suitable to pass through a 500 μm sieve. The rice husk fly ash was supplied by Herba Ricemills S.L., a rice-producing industry in San Juan de Aznalfarache (Seville, Spain). The ash was black because the burning temperature was not controlled. Wood bottom ash was supplied by the cogeneration plant belonging to Tradema S.L., a wooden board factory in Linares (Jaen, Spain). The raw materials are solid by-products generated during the process and composed primarily of wood scraps from conifer bark, wood dust, and non-conforming products. Pine-olive pruning fly and bottom ashes were supplied by the plant Aldebarán Energia del Guadalquivir S.L. located in Andújar (Jaen, Spain). This company generates renewable energy using as fuel biomass of the pruning of olive groves and forest pruning (pine). Olive stone bottom ash was supplied by Valoriza Energía company located in the Scientific and Technological Park Geolit at Mengibar (Jaen, Spain) provided air conditioning to the Park using as fuel olive stone. Olive pomace bottom and fly ash were supplied by the plant Energía de La Loma S.A. located in Villanueva del Arzovispo (Jaen, Spain). This company uses as a fuel, orujillo.

20 wt% of fly or bottom biomass ashes were added to the clay and mixed to obtain good homogenization. For comparative purposes, ten samples per series were prepared for the tests. The necessary amount of water (7-10 wt% moisture) was added to obtain an adequate plasticity and assess the absence of defects, mainly cracks, during the semi-dry compression moulding stage under 54.5 MPa of pressure, using an uniaxial laboratory-type pressing Mega KCK-30 A. Waste-free mixtures were also made as references. Solid bricks with 30 x 10 mm cross sections and a length of 60 mm were obtained. Samples were fired in a laboratory

furnace at a rate of 3ºC/min up to 900ºC for 4 h. Samples were then cooled to room temperature by natural convection inside the furnace. The shaped samples were designated as C for the bricks without waste and C-20W for the mixtures, where W indicated the waste incorporated (RHFA (rice husk fly ash); WBA (wood bottom ash); POPFA (Pine-olive pruning fly ash); POPBA (pine-olive pruning bottom ash); OSBA (olive stone bottom ash); OPBA (olive pomace bottom ash) and OPFA (olive pomace bottom and fly ash)).

2.2. Characterization of Raw Materials

A laser diffractometer Malvern Mastersizer 2000 was used to measure particle size distribution of different ash particles. Crystalline phases were evaluated in a X-ray diffractometry with a X´Pert Pro MPD automated diffractometer (PANanalytical) equipped with Ge (111) primary monochromator, using monochromatic Cu Kα radiation (λ= 1.5406 Å) and an X´Celerator detector. Chemical composition was determined by X-ray fluorescence (XRF) using the Philips Magix Pro (PW-2440). Thermal behavior was determined by thermogravimetric and differential thermal analysis (TGA-DTA) with a Mettler Toledo 851e device in oxygen. The total content of carbon, hydrogen, nitrogen, and sulphur was determined by combustion of samples in O_2 atmosphere using the CHNS-O Thermo Finnigan Elementary Analyzer Flash EA 1112. The organic content was measured according to ASTM D-2974, Standard Test Method for Moisture, Ash, Organic Matter of Peat and Other Organic Soils [33]. The ignition temperature was 440°C. Carbonate content (expressed as calcium carbonate) was determined by calcimetry in a Bernard calcimeter.

2.3. Characterization of Clay-Ash Bricks

Linear shrinkage was obtained by measuring the length of samples before and after the firing stage, using a caliper with a precision of ± 0.01 mm, according to ASTM standard C326 [34]. Water absorption values were determined from weight difference between the fired and water-saturated samples (immersed in boiling water for 2 h), according to ASTM standard C373 [35]. Open porosity (in vol.%) were determined from weight differences between saturated and dry mass with respect to exterior volume; and closed porosity (in vol.%) was calculated from weight difference between dry and suspended mass in water with respect to exterior volume according to ASTM standard C373 [35]. Bulk density was determined by the Archimedes method [35]. Water suction of a brick is the volume of water absorbed during a short partial immersion. Tests to determine water suction was implemented according to standard procedure UNE-EN 772-1 [36]. An efflorescence study was carried out. To this end, the bricks were immersed in water for 24 h and dried in shade. After this treatment the presence of soluble salts in the bricks was determined.

Compressive strength of bricks defined as their bulk unit charge against breakage under axial compressive strength. For this trial, six fired samples were studied. Tests on compressive strength were performed according to standard UNE-EN 772-1 [37] on a MTS 810 Material Testing Systems laboratory press. The area of both bearing surfaces was measured and the average taken. All samples were submitted to a progressively increasing normal strength, with the load applied on the center of the upper surface of the sample until breakage. The compressive strength of each sample was obtained by dividing the maximum load by the average surface of both bearing surfaces, expressed in MPa with 0.1 MPa accuracy.

Table 1. Chemical composition of clay and biomass ash

Oxide content (%)	Clay	RHFA	WBA	POPBA	POPFA	OPBA	OPFA	OSBA
SiO_2	54.4	76.7	48.6	48.1	41.2	19.14	11.7	8.47
Al_2O_3	12.36	0.183	5.94	9.3	8.05	3.47	2.51	1.68
Fe_2O_3	4.58	0.233	3.26	3.2	2.78	1.62	1.26	2.97
CaO	8.76	0.821	18.1	17.7	22.1	15.00	10.2	24.00
MgO	2.46	0.654	3.2	2.73	4.63	4.43	3.03	3.42
MnO	0.03	-	0.051	0.10	0.178	0.039	-	0.057
Na_2O	-	-	0.92	-	-	-	-	-
K_2O	3.37	2.03	1.85	3.83	6.78	28.32	42.66	31.22
TiO_2	0.60	-	1.39	0.473	0.467	0.158	0.11	0.073
P_2O_5	0.11	1.62	0.52	1.01	1.99	4.11	2.97	4.04
SO_3	0.68	-	0.14	0.023	0.28	0.86	3.6	-
ZnO	0.026	-	0.281	0.023	0.065	0.023	-	0.025
SrO	0.027	0.035	0.0434	0.047	0.047	0.056	0.037	0.148
ZrO_2	0.033	-		-	-	0.012	-	-
Cl	-	-	0.064	-	0.33	0.64	2.26	0.074
LOI	12.51	17.78	15.62	12.73	9.95	22.09	18.54	23.79

The development of porosity in samples was also evaluated by means of a scanning electron microscope (SEM), using the high-resolution transmission electron microscope JEOL SM 840, and the Energy Dispersive X-ray

Spectroscopy (EDS) for chemical analysis (20 kV). Samples were placed on an aluminium grate and coated with carbon using the ion sputtering device JEOL JFC 1100.

Leachability of heavy metals in the samples was studied using the toxicity characteristic leaching procedure (TCLP) according to EPA method 1311 [38]. The concentrations in the filtrate were measured with an Inductively Coupled Plasma-Atomic Emission Spectrometer (ICP-AES Agilent 7500).

3. Results and Discussion

3.1. Characterization of Raw Materials

Table 1 shows the chemical composition of raw clay and different biomass ashes, determined by XRF. The raw clay had high amounts of SiO_2 and Al_2O_3 as the predominant oxides, due mainly to the silicate in the clay. The content of CaO was significant and it is related to the abundance of carbonates. It had a relatively high content of Fe_2O_3, which confers the red colour. The rice husk ash contains a high proportion of pure silica (79.6%). The main components observed in the wood ash, and pine-olive pruning fly and bottom ashes were SiO_2, CaO and Al_2O_3. The minor components were Fe_2O_3, K_2O and MgO. On the other hand, K_2O is the major oxide present followed by CaO and SiO_2 in olive stone bottom ash and olive pomace bottom and fly ashes. High content in K_2O is due to the oxide being a major component of the olive stone and pomace used as fuels. Chemical components of ceramic bodies can be classified into three groups: (1) skeleton components, which form the framework and surface of ceramic bodies, mainly consisting of SiO_2 and Al_2O_3; (2) flux materials, which lowered the melting point, mainly containing alkaline metal oxide and alkaline-earth metal oxides such as CaO, Na_2O, K_2O and MgO; (3) gaseous components, which generate gases and bloat ceramic bodies during sintering at high temperature [39]. As indicated in Table 1, the total content of Si and Al (estimated as oxides) is higher in rise hush fly ash, wood bottom ash, pine-olive pruning fly and bottom ashes. High contents of fluxing oxides (K_2O) and auxiliary fluxing oxides (CaO, MgO) in olive stone bottom ash and olive pomace bottom and fly ashes are suitable to lower the temperature of the firing process for brick preparation.

The XRD patter of raw clay showed SiO_2, $CaCO_3$ as main crystalline phases and alumina-silicates of K-Mg or Fe-Mg in smaller proportion. X-ray diffraction patterns of the biomass ashes showed a series of crystalline phases within a glassy matrix. In the X-ray diffraction pattern of WBA, POPBA and POPFA the major phases identified were SiO_2 and $CaCO_3$. The presence of tiny diffraction peaks indicated the presence of crystalline silicates of Al and Fe with alkaline earth elements ($AlKSi_3O_8$ and $Ca_{5.74}Fe_{0.26}Si_6O_{18}$) in WBA and some aluminosilicates such as $Al_{0.5}Ca_2Mg_{0.75}O_7Si_{1.75}$, $Al_{0.83}Ca_{3.027}Fe_{1.17}O_{12}Si_3$ and silicoaluminatesof Na and K in POPBA and POPFA.

XRD patterns of OSBA, OPBA and OPFA are also complex and presenting a great number of diffraction peaks. These samples contain great quantities of potassium as observed from XRF analysis. Several diffraction peaks associated to K-crystalline phases were noticeable in all cases. As before, these samples are composed mainly of silicate, aluminosilicates of earth and alcalino-earth metals as well as carbonates. Some diffraction peaks were difficult to assign but pointed to the presence of phosphate and pyrophosphate compounds.

Table 2. Carbonate content and CNHS analysis of raw materials

Sample	**Organic matter content (%)**	**Carbonate content[a] (%)**	**%C**	**% H**	**%N**	**%S**
Clay	2.29 ± 0.09	7.36 ± 0.33	2.25 ± 0.01	0.34 ± 0.004	0.05 ± 0.002	0.032 ± 0.008
RHFA	16.37 ± 0.17	5.61 ± 0.55	11.45 ± 0.04	0.248 ± 0.005	0.127 ± 0.0	0.0
WBA	10.41 ± 0.09	17.25 ± 0.76	6.15 ± 0.03	0.622 ± 0.021	0.204 ± 0.006	0.025 ± 0.001
POPBA	5.10 ± 0.10	11.41 ± 0.10	9.14 ± 0.18	0.27 ± 0.022	0.04 ± 0.006	0.0
POPFA	3.00 ± 0.08	18.95 ± 0.56	4.79 ± 0.02	0.05 ± 0.019	0.04 ± 0.003	0.018 ± 0.002
OPBA	7.80 ± 0.18	30.35 ± 0.75	4.29 ± 0.01	0.64 ± 0.005	0.030 ± 0.00	0.054 ± 0.009
OPFA	4.70 ± 0.27	31.20 ± 0.74	3.34 ± 0.06	0.47 ± 0.01	0.043 ± 0.00	0.211 ± 0.016
OSBA	11.83 ± 0.12	39.54 ± 0.89	11.60 ± 0.01	0.40 ± 0.02	0.012 ± 0.0	0.0

[a]Determined according ASTM D-2974.

Oxides, such as quartz (SiO_2) are produced at high temperature during the combustion process. Carbonates were also detected due to biomass fuel contains a high content of wood waste [40]. The detection of silica and calcite depends on the combustion temperature and the type of biomass used as fuel [41]. Traces of clay minerals were also detected. The mineralogical composition obtained for biomass ashes are consistent with other studies [42-44]. Also, according with background intensity, high content of glass phases indicated high activity of biomass ash to reduce the sintering temperature in the heating procedure. The average particle size, D_{50}, was lower for POPFA and OPFA (fly ash), however RFHA presented an average particle size higher than the bottom ashes (Table 3). Therefore, POPHFA and OPFA have a higher reaction area per unit volume and higher surface energy per area, thus resulting in a higher activity [45].

Table 3. Average particle size D_{50} for fly and bottom biomass ashes

Biomass ash	Average particle size D_{50} (μm)
RHFA	96,13
WBA	92,14
POPBA	58,68
POPFA	17,41
OPBA	61,77
OPFA	17,43
OEBA	59,44

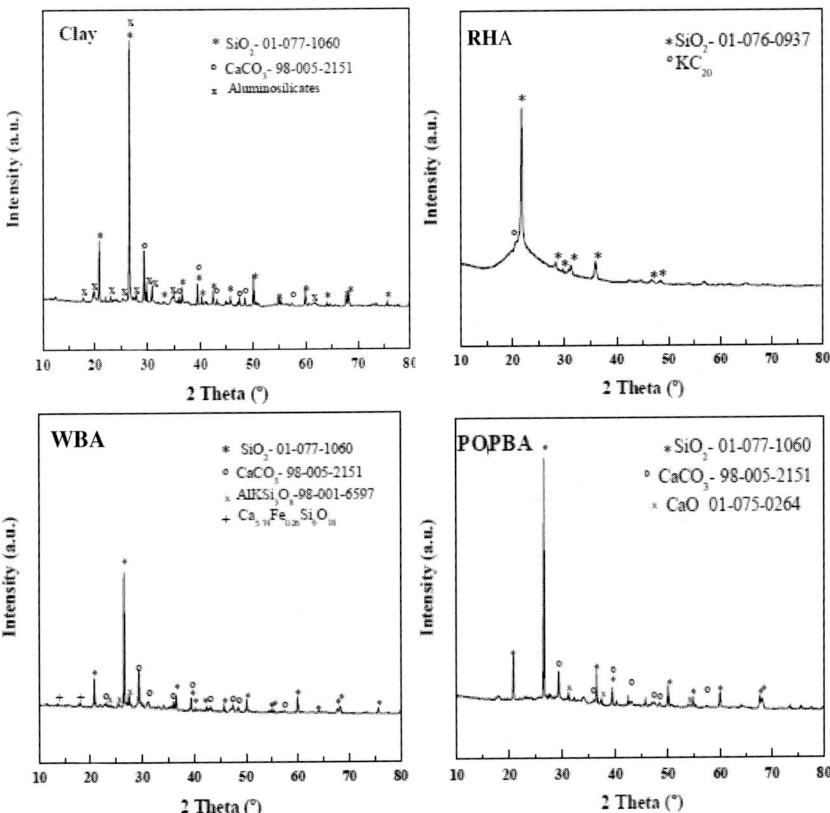

Figure 1. Continued on next page.

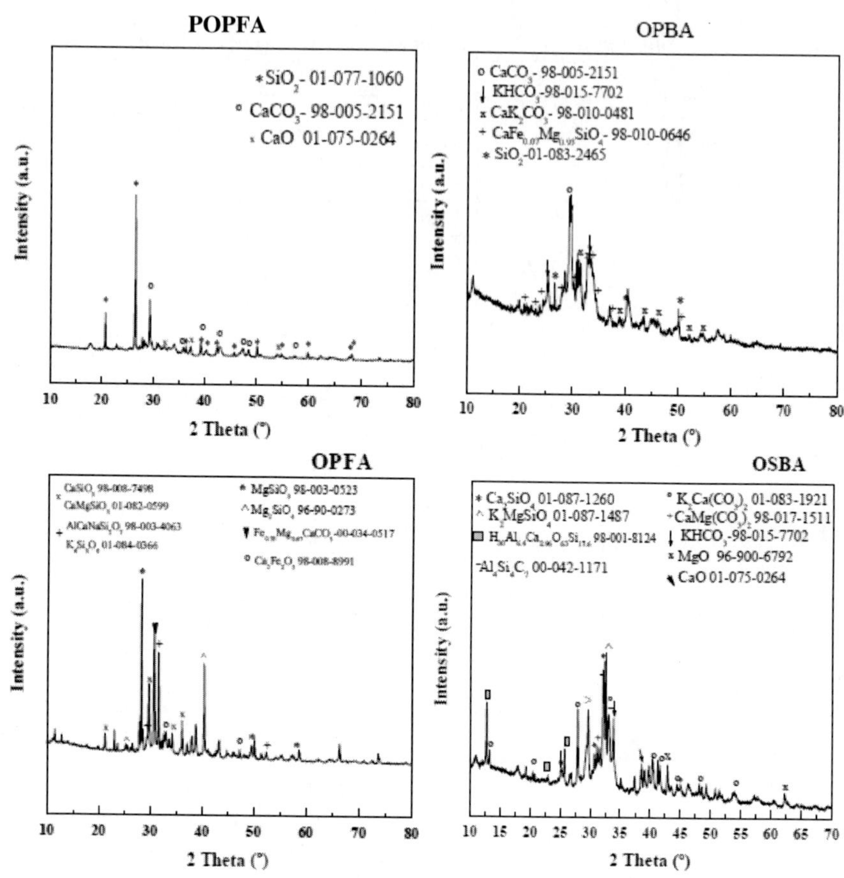

Figure 1. XRD patterns of clay; RHFA; WBA; POPBA; POPFA; OPBA; OPFA and OSBA.

The CNHS analysis of biomass ashes (Table 2) showed that biomass ashes were composed mainly of carbon, hydrogen and a small quantity of nitrogen. OPFA, OPBA, WBA and POPFA also contained a small amount of sulphur. The carbon content is equivalent to the inorganic carbon derived from the decomposition of carbonates. Carbon content in ash usually depends on the efficiency of combustion technology. Large amounts of organic carbon in ash indicate incomplete combustion of biomass, which suggests inefficient fuel use [46].

The TGA/DTA curves of the clay and biomass ashes up to 900°C are shown in Figure 2. The weight loss of 1.0% between 30-150ºC of the raw clay, associated to the release of hydration water, showed an endothermic peak centered at 70°C. The weight loss of 2.2% between 150°C and 600°C can be attributed to the combustion of organic matter as indicates the exothermic effects centered at

375 and 475ºC and to the silicate dehydroxylation as indicated the endothermic peak centered at 570ºC. Between 600°C and 800°C, the decomposition of calcium carbonate with the release of CO_2 was indicated by the endothermic effect centered at 760°C with a mass loss of 8.2%.

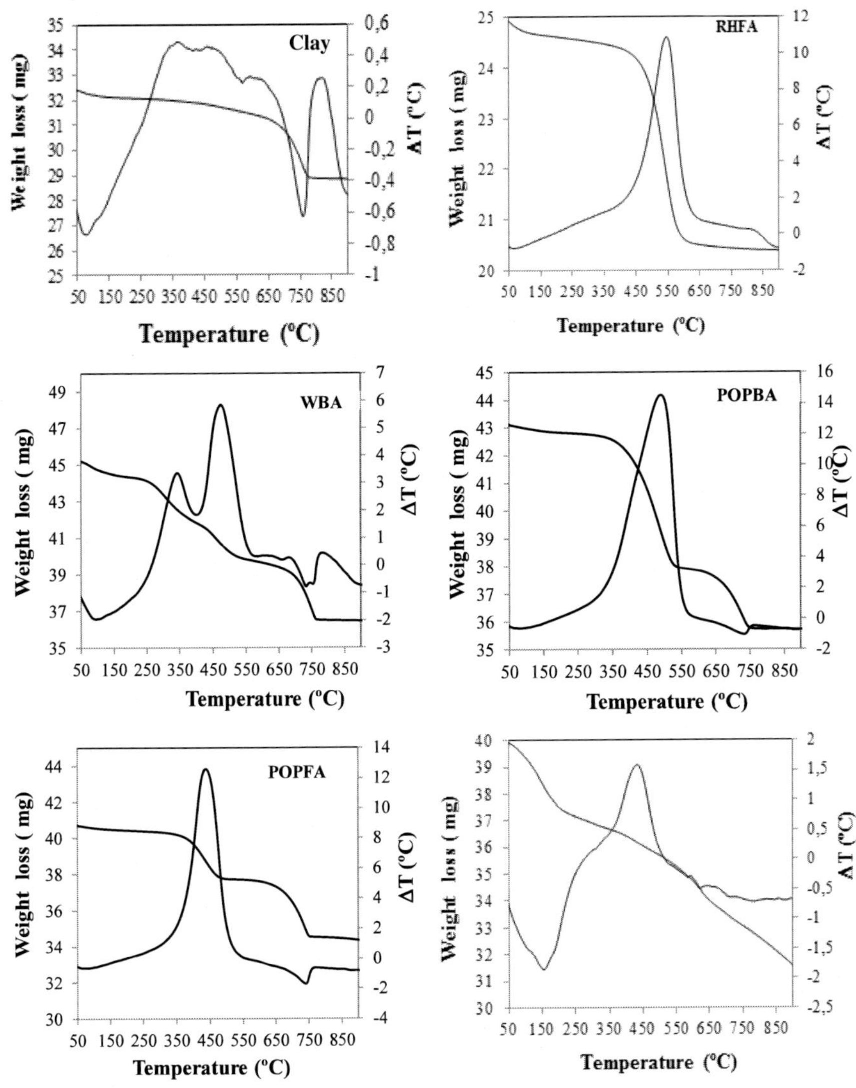

Figure 2. Continued on next page.

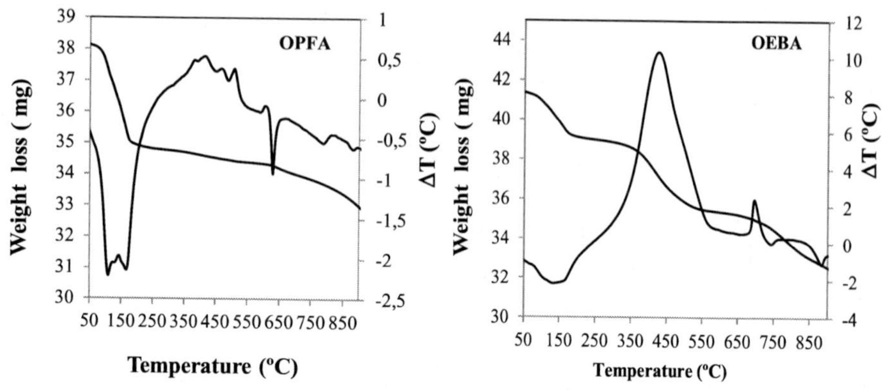

Figure 2. DTA/TGA analysis of clay; RHFA; WBA; POPBA; POPFA; OPBA; OPFA and OSBA.

The TGA/DTA results of seven biomass ashes (Figure 2) indicated that weight loss curves are complex with gradual weight losses up to the final temperature 900ºC. The first weight loss at low temperature (30-200ºC) is due to, as indicated the endothermic peak, the presence of moisture in the sample. Release of organic matter (exothermic peak) and unburned elements (exothermic peak), such as residual carbon, occurred from 200-650ºC, indicating, in all cases, unburned carbon in the ash. Between 650-900ºC the decomposition of calcium carbonates (endothermic peak) emitting CO_2 takes place. In all the biomass ashes except in the RHBA samples, several thermal effects occurred at higher temperatures. At 650ºC, an endothermic peak was observed due to elimination of structural water from the hydroxide ions in the biomass ash. The exothermic peak at 700ºC was probably ascribed to combustion reactions of the unburned elements in the ashes, while the endothermic effect at 750ºC may be due to carbonates decomposition.

3.2. Characterization of the Clay-Biomass Ash Bricks

During firing process, changes of dimensions, mass and colour variations occurs into the clay-biomass ash bricks. It was observed that the bricks became darker, after firing at 900ºC the colour is reddish due to the iron content in clayed materials. No defects such as cracks, bloating or efflorescence were found after firing. Different technological properties of the biomass ash-clay bricks compared to the standard (only clay mixture) were studied.

The linear shrinkage of the bricks reflects expansion/contraction behavior during thermal treatment. Shrinkage is a key parameter to evaluate the quality of fired bricks. The stress on the clay body increases as the shrinkage so does, therefore, large shrinkage may induce to tensions and even to broken pieces [47]. As indicated in Figure 3 addition of the biomass ash changed the linear shrinkage of the clay. The standard bricks had a linear shrinkage of -0.52 showing expansion behavior. Only the addition of RHFA produced contraction in the bricks. However, the addition the OSBA produced the highest increase in the linear shrinkage up to -1.88%. The addition of POPFA, OPFA and OPBA also produced an increased in this property showing values of -0.65%, -1.23% and -1.62% respectively. However the addition of POPBA waste does not change the linear shrinkage of the clay. Thus, clay samples, as well as samples containing biomass ash, except RHFA-clay bricks, expanded slightly when fired at 900°C. This behavior could be due to the expansion in the ceramic body of the biomass ash amended bricks as a result of the gases released in the combustion of organic matter and carbonates decomposition, which caused the expansion. The addition of RHFA could reduce, at 900°C, the liquid phase viscosity and the gases resulting from organic matter decomposition and $CaCO_3$ could escape, producing shrinkage in the bricks. The values of linear shrinkage suggest that the introduction of 20 wt% of biomass ash showed linear shrinkage lower than 2.0%, within the safety limits for industrial production of clay bricks.

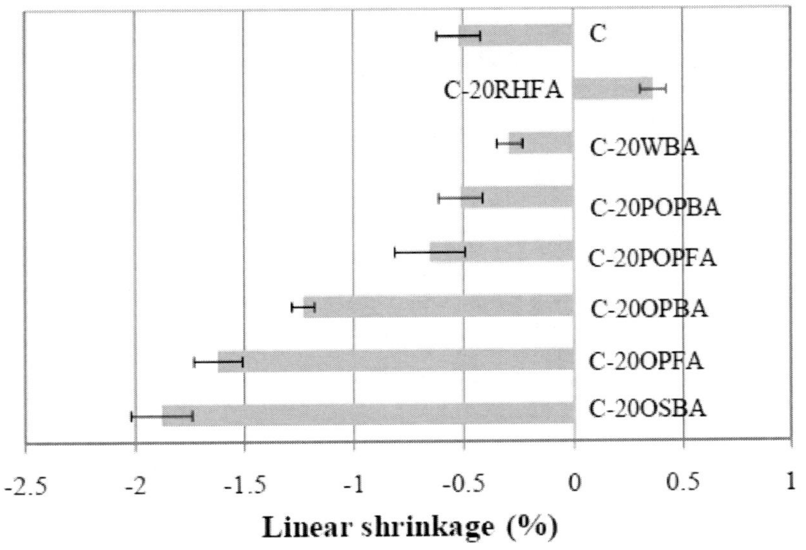

Figure 3. Linear shrinkage of clay bricks and clay–biomass ash bricks.

Figure 4 shows the weight loss of ignition of the brick samples. The control clay bricks present the lowest loss on ignition (9.2%). The addition of biomass ash increased the loss of ignition to values varying from 12% up to 15.2% with the addition of POPFA and OSBA respectively. The increase in the weight loss on ignition during the firing process could increase due to the contribution of organic matter and carbonate content as well as inorganic matter in clay and biomass ashes.

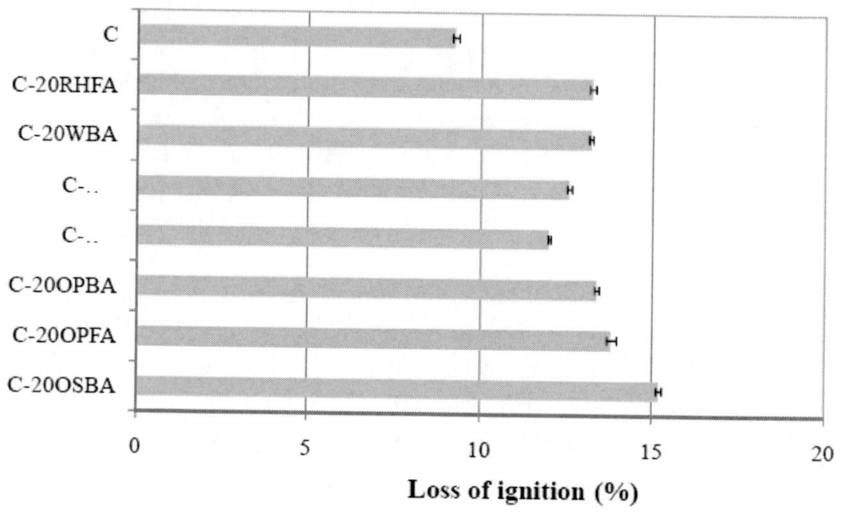

Figure 4. Loss of ignition of clay bricks and clay-biomass ash bricks.

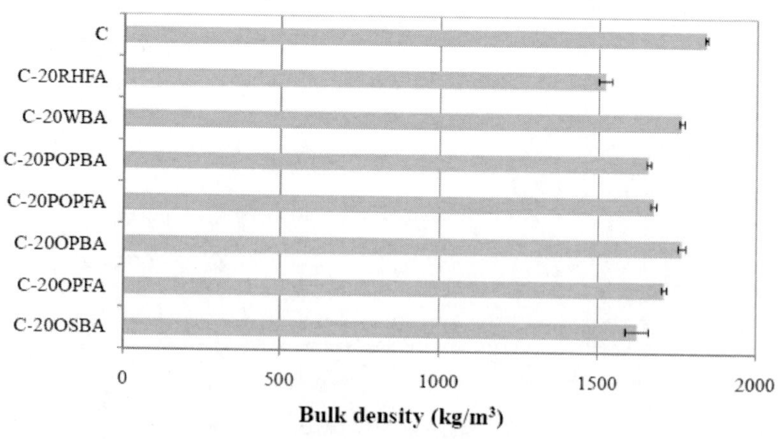

Figure 5. Bulk density of the fired bricks as function of type of biomass ash addition.

The addition of biomass ashes decreased the bulk densities of the fired bricks (Figure 5), which exhibited values ranging from 1,765to 1,520 kg/m^3 lower than the values for control bricks of 1,832 kg/m^3. The addition of 20 wt% of biomass ash decreased the bulk density in this order: WBA=OPFA>OPBA>POPFA>POPBA>OSBA>RHFA. The reduction in bulk density indicated that this type of wastes promotes an expansion reaction at firing temperature of 900ºC. In addition, the porosity increases with biomass ashes addition also playing an important role in reducing the weight of the bricks, declining labor and transportation costs.

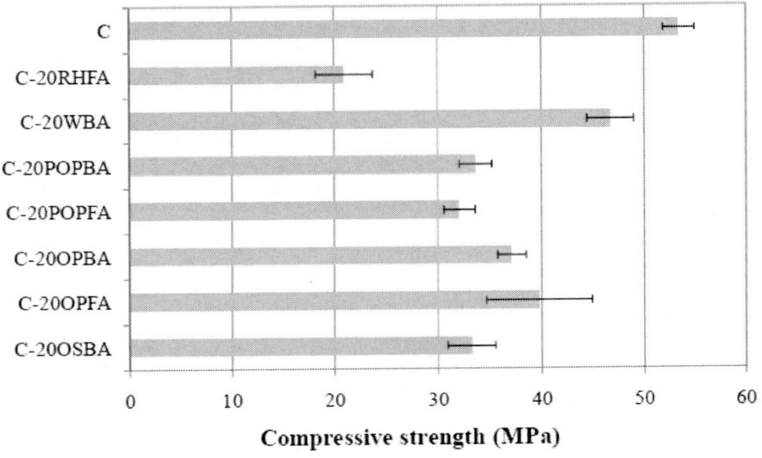

Figure 6. Compressive strength of clay bricks and clay-biomass ash bricks.

The compressive strength is another important parameter to consider in order ensuring the engineering quality of the bricks. The results of compressive strength are shown in Figure 6. Test results showed that the standard bricks had higher compressive strength (53.4 MPa). The highest compressive strength is achieved with the addition of WBA, 46.7 MPa, a reduction of 12.5% compared to the bare clay. If OPFA and OPBA are added, the compressive strength is reduced by 25-30%, respectively; and 37-39% with the addition of OSBA, POPBA and POPFA. The highest reduction in the compressive strength with respect to control bricks is obtained with the addition of RHFA, this parameter decreased up to a 60.8%. It should be pointed that the compressive strength of bricks mainly depends on their bulk density, amount and type of porosity, and pore size distributions [48-50]. A linear relationship between compressive strength and apparent porosity as well as compressive strength and water absorption was observed as shown in Figure 7.

Decomposition of organic matter and carbonates can lead to higher porosity. Samples containing biomass ash had lower bulk density and greater total porosity, having higher open porosity than control brick based on water absorption data. Open pores and other microscopic imperfections acted as stress concentrator notches and reduced the compressive strength. The high amount of silica present in RHFA can lead to higher porosity. Generally, silica content in clay bricks ranged between 50-60%, higher amount is undesirable [51], since reduces the mechanical properties of bricks. RHFA has a 76.7% of silica and explains the lower bulk density and higher apparent porosity and water absorption. According to ASTM C62-10 [52] and European Standard EN-772-1 [37], the compressive strength varies from 10 MPa for weather-resistant brick to 20 MPa in the case of severe weathering. Therefore, it can be used 20 wt% of biomass ashes to replace clay for sustainable bricks. Only samples with 20 wt% of RHFA content had compressive strength rates close to the minimum (20 MPa) required by the standard threshold value for severe weathering.

Apparent porosity as a function of biomass ashes is shown in Figure 8. It can be observed that standard bricks have the lowest apparent porosity (31.0%) whereas RHFA bricks have highest apparent porosity values (43.0%). The addition of WBA and OPFA increased the apparent porosity by 10%, 17% for OPBA and OSBA, 25% for POPFA and POPBA and 39% for RHFA. Porosity is related to biomass residual combustion, carbonate decomposition and dehydroxilation reactions [53].

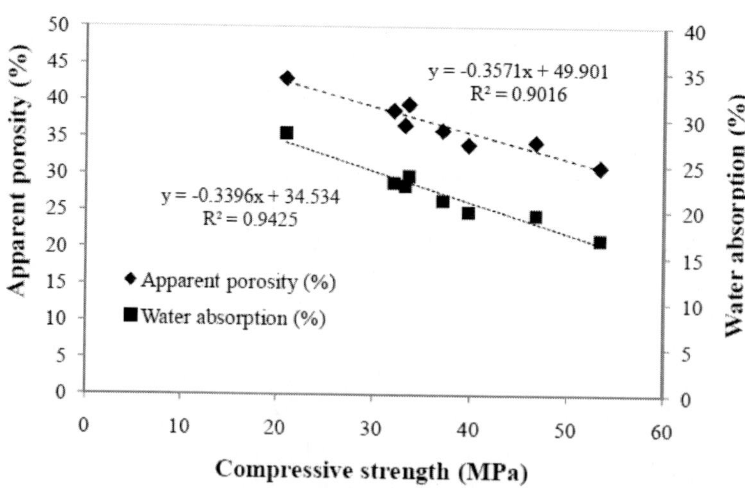

Figure 7. Relationship between compressive strength and apparent porosity and compressive strength and water absorption of clay-biomass ash bricks.

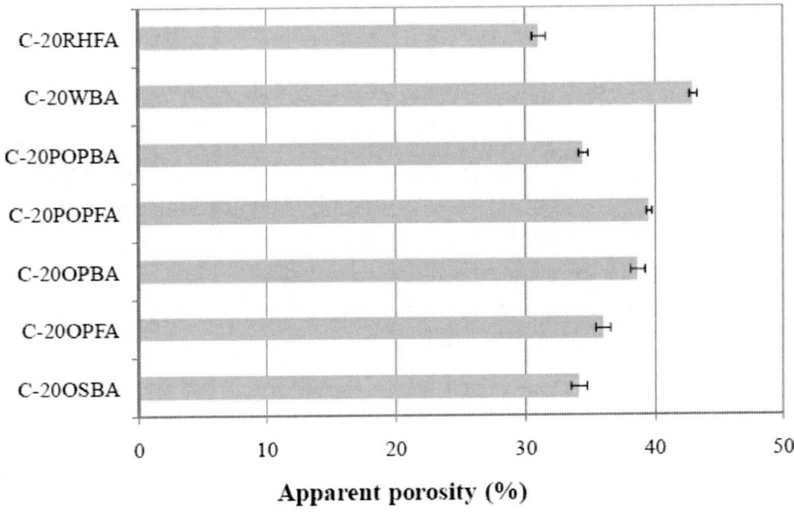

Figure 8. Apparent porosity of clay bricks and clay-biomass ash bricks.

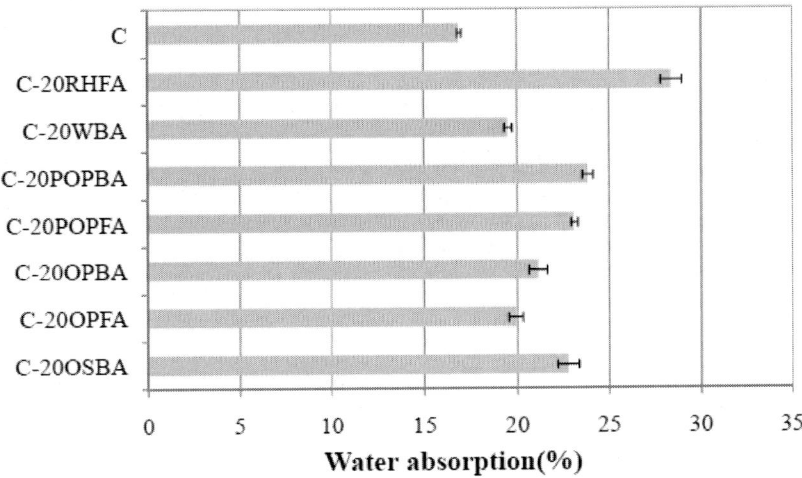

Figure 9. Effect of biomass ash addition on water absorption of the fired bricks.

Water absorption is an important factor affecting the durability of bricks. Low water absorption values indicate a greater life of ceramic specimens and resistance to external weather conditions. The test results of the water absorption increased with biomass ash incorporated in the bricks (Figure 9). Control bricks showed the minimum water absorption of 16.9% since they had the highest density and the lowest apparent porosity. It is noteworthy that C-20RHBA bricks showed the maximum water absorption value (28.4%), while the minimum water absorption value is obtained by incorporating WBA (19.5%). The open porosity increased in the following order: WBA<OPFA<OPBA<OSBA<POPFA<POPBA<RHFA. A linear relationship between apparent porosity and water absorption was observed (Figure 9). Water absorption of bricks is classified in accordance to the different grades of weathering exposure condition according to ASTMC67-07a:2003 [54]. For severe weathering resistance bricks, water absorption may not be higher than 17%. For moderate weathering resistance bricks, water absorption will not be higher than 22% and no limit is set for negligible weathering resistance bricks. Water absorption of WBA, OPFA and OPBA bricks were lower than 22%, so they can be used in moderate weather, however water absorption of OSBA, POPFA, POPBA and RHFA were higher than 22%, what could cause various kinds of damage to the bricks as appearance of cracks with the consequent structural damage in the building [55].

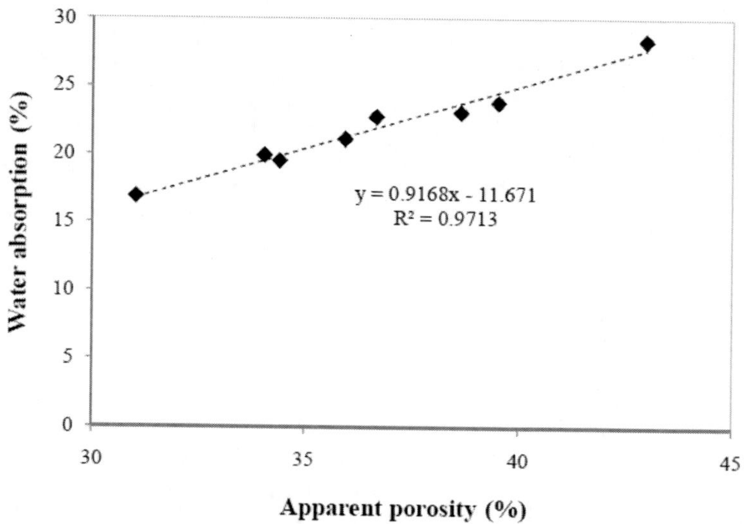

Figure 10. Relationship between apparent porosity and water absorption of clay-biomass ash bricks.

Type of biomass ashes incorporated to the clay affected the water suction. Brick and mortar bond is influenced by the water suction and water retaining of mortar. The brick should not absorb a large amount of water to the mortar has enough water of hydration and harden at a speed that allows the correct adhesion with the bricks. Figure 10 shows the water absorption for bricks containing different biomass ashes. The addition of biomass ashes produced a significant increase in suction water, for example control bricks show the minimum suction water value 2.3 kg/m^{2}min. The incorporating of 20 wt% of RHFA to clay showed the maximum suction water of 4.3 kg/m^{2}min, while the incorporation of 20 wt% of OPFA and WBA produced the minimum suction water around 2.8 kg/m^{2}min. This can be due to a raise in the interconnected surface porosity under the presence of biomass ashes, possibly due to a growth in pores, both in size and in number, caused by the organic waste and carbonate content. For all the tested specimens, the water absorption was lower than the higher limit 4.5 kg/m^2 min [56], meaning that all the biomass ash-containing bricks comply with this criterion, indicating an efficient bond between brick and mortar.

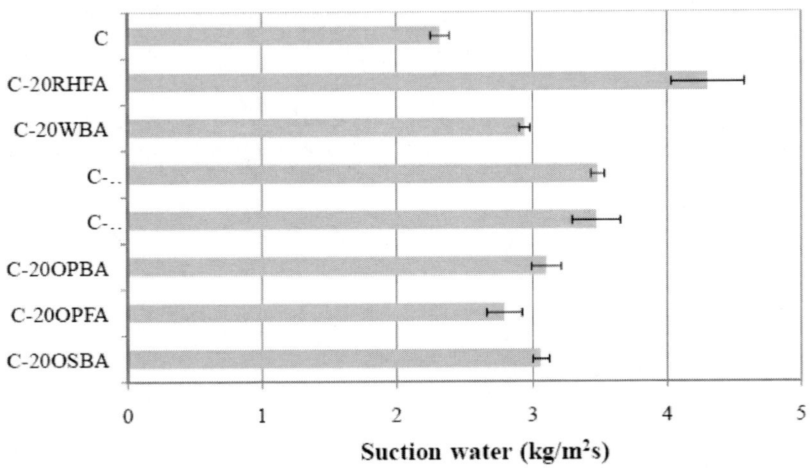

Figure 11. Water suction of clay bricks and clay-biomass ash bricks.

The evolution of the porosity with the type of biomass ash content were also analyzed using SEM. Figure 11 shows the SEM observations for clay bricks and some representative bricks as C-20OPFA, C-20RHFA and C-20POPBA. Clay porosity is mainly open porosity with a lower proportion of closed pores. Addition of biomass ash changed the clay porosity. RHFA waste additions increased the porosity of clay, producing primarily open porosity, as well as, macropores, according to water absorption data. Clay-RHFA micrographs showed some tiny spheres, known as cenospheres, that are perfectly round. The addition of OPFA and POPBA showed the presence of less densification where interconnected irregular shaped open pores were observed. This fact was more important in the case of RHFA containing bricks in accordance to bulk density and water absorption data. The increase in the porosity under the presence of biomass ashes provokes a reduction of the mechanical properties of ceramic bricks, however, it may help to strengthen insulation in building bricks, since thermal conductivity is highly influenced by porosity.

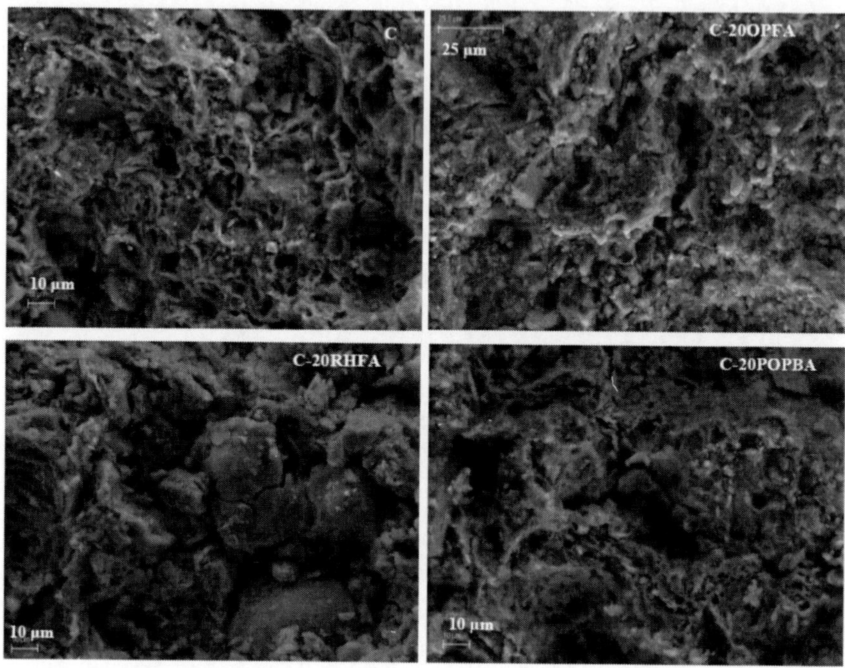

Figure 12. SEM micrographs of clay; C-20OPFA; C-20 RHFA; C-20POPBA.

Finally, the environmental effect of using biomass ash in clay bricks has been investigated. Heavy metals in the environment can have harmful effects on human and animal health. In order to assess the environmental impact of the biomass ash bricks, the leaching of heavy metals is an important index to evaluate the adsorption behavior and the effect of immobilization. Table 4 shows the results of the test of leaching of heavy metals from the control bricks reference and biomass ash-clay bricks. In general, it can be observed that the use of biomass ashes gives rise to concentrations of heavy metals that are far below the limits established by EPA 658/2009 [38] and Spanish legislation (MO AAA/661/2013) [57] on disposal of hazardous waste. These biomass ashes can be classified as acceptable in landfills or inert and non-hazardous waste. Therefore, leaching tests indicated a high degree of immobilization of heavy metals, indicating that the incorporation of the different types of biomass ashes in the ceramic products is an efficient method of inerting.

Table 4. USEPA TCLP test results (ppm) and the maximum concentration of contaminants for toxicity characteristics

Component (ppm)	Sample								USEPAregulated TCLP limits (ppm)
	C	**C-20RHFA**	**C-20WBA**	**C-20POPBA**	**C-20POPFA**	**C-20OPBA**	**C-20OPFA**	**C-20OSBA**	
As	0.027	0.141	0.050	0.085	0.101	0.017	0.013	0.028	5
Ba	0.204	0.160	0.763	0.224	0.203	0.562	0.800	0.558	100
Cd	0.0003	0.00114	0.00041	0.117	0.0017	0.0001	0.0002	0.00012	1
Co	0.010	0.049	0.059	0.0145	0.0197	0.0114	0.0263	0.0013	-
Cr	0.554	0.131	0.138	0.145	0.182	0.254	0.109	0.282	5
Cu	0.296	0.294	0.533	0.565	1.165	1.317	0.694	0.986	5
Ni	0.047	0.063	0.106	0.037	0.032	0.051	0.046	0.012	-
Pb	0.0009	0.00258	0.0010	0.0023	0.259	0.081	0.288	0.087	5
Sb	0.0010	0.0013	0.0036	0.0027	0.0007	0.00044	0.0004	0.0098	-
Se	0.013	0.010	0.010	0.0095	0.008	0.0049	0.0050	0.0031	1
Sn	0.00014	0.00006	0.00096	0.02	0.00008	0.0002	0.00017	0.00016	-
V	0.161	0.181	0.255	0.192	0.154	0.176	0.107	0.200	-
Hg	0.00004	0.00007	0.00005	0.000058	0.00004	0.00012	0.000052	0.00011	0.2
Zn	0.058	0.010	0.236	0.086	0.667	0.079	0.106	0.005	300

Conclusion

This study investigated the possible incorporation of different biomass ashes in fired clay bricks. Seven biomass ash wastes (RHFA, WBA, POPBA, POPFA, OPBA, OPFA and OSBA) generated in different plant in Andalusia (Spain) were valued. Bricks incorporating 20 wt% of biomass ash and standard bricks with 0% of waste were investigated by their physical and mechanical properties. According to the test results it was found that:

1. The main component of RHFA is mainly SiO_2. The WBA and POPBA and POPFA are mainly composed by SiO_2, CaO and Al_2O_3, while K_2O is the major oxide present followed by CaO and SiO_2 in OPBA, OPFA and OSBA. The XRD analysis proved the presence of significant amounts of amorphous materials. The composition is similar to the clay used as raw material.

2. The compressive strength test indicated that mechanical strength of bricks decreased between 12-40% with the addition of biomass ash, being the most pronounced decline (up to 61%) with the addition of RHFA, due to its high content of silica affecting the compressive strength of bricks. However all biomass ash-bricks satisfy the requirement of compressive strength specified by ASTM C62-10 and European Standard EN-772-1 for masonry construction materials.
3. The bulk density of the brick specimens were reduced by 4.0-17.3% after incorporating biomass ashes. However the increase of apparent porosity lead to an important increased of water absorption. Only WBA, OPBA and OPFA showed water absorption values lower than 22%. These bricks with high porosity usually have good insulation properties.
4. SEM micrographs corroborate the increase in porosity with the incorporation of biomass ashes associated with the amount of organic content and carbonate content in the raw materials.
5. The TCLP leaching concentrations for the target metals in the control clay bricks and biomass ash-clay bricks met the current regulatory thresholds for the EPA and for Spain.
6. The incorporation of biomass ash in the brick production can be an economical and sustainable solution for the construction sector.

Acknowledgments

This work has been funded by the Project "Valuation of various types of ash for the obtaining of new sustainable ceramic materials" (UJA2014/06/13), Own Plan University of Jaen, sponsored by Caja Rural of Jaen. Technical and human support provided by CICT of Universidad de Jaén (UJA, MINECO, Junta de Andalucía, FEDER) is gratefully acknowledged.

References

[1] Muneer, T., Asif, M., Munawwar, S., 2005. Sustainable production of solar energy with particular reference to the Indian economy. *Renew. Sust. Energ. Rev.* 9, 444-473.

[2] Dincer, I., 2000. Renewable energy and sustainable development: a crucial review. *Renew. Sust. Energ. Rev.* 4, 157-175.

[3] Acikgoz, C., 2011. Renewable energy education in Turkey *Renew. Energy* 36, 608-611.

[4] DIRECTIVE 2003/54/EC.

[5] Plan de Acción Nacional de Energías Renovables (PANER) 2011-2020 [[National Action Plan for Renewable Energy (NREAP) 2011-2020]. Ministerio de Energía Industria y Turismo. Gobierno de España [Ministry of Energy, Industry and Tourism. Government of Spain]. http://www.minetur.gob.es/energia/desarrollo/EnergiaRenovable/Paginas/Paner.aspx.

[6] Ericsson E, 2007. Co-firing – a strategy for bioenergy in Poland. *Energy* 32, 1838-1847.

[7] Andalusian Energy Agency (AEA) [Agencia Andaluza de la Energía]. 2015, La biomasa en Andalucía [Biomass in Andalusia]. Consejería de Empleo, Empresa y Comercio [Department of Employment, Business and Trade]. Junta de Andalucía.

[8] García-Maraver, A., Zamorano, M., Ramos-Ridao, A., Díaz, L. F. M., 2012. Analysis of olive grove residual biomass potential for electric and thermal energy generation in Andalusia (Spain). *Renew. Sust. Energ. Rev.* 6, 745-751.

[9] Food and Agriculture Organization of the United Nations – Statistics Division, 2016. <http://faostat3.fao.org> (last visited 28 june 2016).

[10] Caputo, A.C., Scacchia, F., Pelagagge, P.M., 2007.Disposal of by-products in olive oil industry: waste-to-energy solutions. *Appl. Therm. Eng.* 23, 197-214.

[11] Ramachandran, S., Singh, S.K., Larroche, C., Soccol, C.R., Pandey, A., 2007. Oil cakes and their biotechnological applications – a review. *Bioresour. Technol.* 2007, 98, 2000-2009.

[12] Arvanitoyiannis, I.S., Kassaveti, A., 2008.Olive oil waste management: treatment methods and potential uses of treated waste. *Waste Manage. Food Industries* 8, 453-568.

[13] Abu Tayeh, H., Najami, N., Dosoretz, C., Tafesh, A., Azaizeh, H., 2014.Potential of bioethanol production from olive mill solid wastes. *Bioresour. Technol.* 152, 24-30.

[14] Kinab, E.,Khouri, G., 2015. Management of olive solid waste in Lebanon: from mill to stove. *Renew. Sustain. Energy Rev.* 52, 209-216.

[15] Christoforou, E., Fokaides, P.A., 2016. A review of olive mill solid wastes to energy utilization techniques. *WasteManage.* 49,346-363.

[16] Cabrera, M., Galvin, A.P., Agrela, F., Carvajal, M.D., Ayuso, J., 2014. Characterisation and technical feasibility of using biomass bottom ash for civil infrastructures. *Constr. Build. Mater.* 58, 234-244.

[17] Rajamma, R., Ball, R, Tarelho, L., Allen, G., Labrincha, J., Ferreira, V., 2009.Characterization and use of biomass fly ash in cement-based materials. *J. Hazard. Mater.*2009, 172 (2-3), 1049-1960.

[18] Cuenca, J., Rodriguez, J., Martin-Morales, M., Sanchez-Roldan, Z., Zamora, M., 2013.Effects of olive residue biomass fly ash as filler in self-compacting concrete. *Constr. Build. Mater.* 40, 702-709.

[19] Ramchandra, P., 2016. Potential applications of rice husk ash waste from rice husk biomass power plant. *Renew. Sust. Energ. Rev.* 53, 1468-1485.

[20] Zhang, L., 2014. Production of bricks from waste materials – a review. *Constr. Build. Mater.* 47, 643-655.

[21] Leiva, C., Arenas, C., Alonso-Fariñas, B., Vilches, L.F., Peceño, B., Rodríguez-Galán, M., Baena, F. J., 2016. Characteristics of fired bricks with co-combustion fly ashes. *Build Engineer.* 5, 114-118.

[22] Rauta S., Ralegaonkara R., Mandavganec S., 2013. Utilization of recycle paper mill residue and rice husk ash in production of light weight bricks. *Arch. Civ. Mechan. Eng.* 13, 269-275.

[23] Haiying, Z., Youcai, Z., Jingyu, Q., 2011. Utilization of municipal solid waste incineration (MSWI) fly ash in ceramic brick: Product characterization and environmental toxicity.*Waste Manage.* 31, 331-341.

[24] Çiçek, T., Çinçin Y., 2015. Use of fly ash in production of light-weight building bricks. *Constr. Build. Mater.* 94, 521-527.

[25] Sivakumar, N., Almamon Yousef, O. M.,Kamal Nasharuddin, M., 2015. Performance of bricks made using fly ash and bottom ash. *Constr. Build. Mater.* 96, 576-580.

[26] Vilches, L.F., Leiva, C., Vale, J., Fernández-Pereira, C., 2005. Insulating capacity of fly ash pastes used for passive protection against fire. *Cem. Concr. Compos.* 27, 776-781.

[27] Leiva, C.; Vilches L.F.; Fernández-Pereira, C.; Vale, J. 2005. Influence of the type of ash on the fire resistance characteristics of ash-enriched mortars. *Fuel* 84, 1433-1439.

[28] Vilches, L.F., Fernández-Pereira, C., Olivares del Valle, J., Rodriguez Piñero, M.A., Vale, J., 2002. Development of new fire-proof products made from coal fly ash: the CEFYR project.*J. Chem. Technol. Biotechnol.* 77,361-366.

[29] Vilches, L.F., Fernández-Pereira, C., Olivares del Valle, J.,Vale, J., Recycling potencial of coal fly ash and titanium waste as new fire-proof products. *Chem. Eng. J.* 95, 155-161.

[30] Eliche-Quesada, D.; Leite-Costa, J., 2016. Use of bottom ash from olive pomace combustion in the production of eco-friendly fired clay bricks. *Waste Manage*. 48, 323-333.

[31] Cheeseman, C. R., Sollars, C.J, McEntee, S., 2003. Properties, microstructure and leaching of sintered sewage sludge ash. *Resour. Conserv. Recy.* 40, 13-25.

[32] Kazmi, S.M.S., Safeer A.; M., Saleem M.A., Munir, M.J., Khitab, A., 2016. Manufacturing of sustainable clay bricks: Utilization of waste sugarcane bagasse and rice husk ashes. *Constr. Build. Mater.* 2016, 120, 29-41.

[33] ASTM D-2974:1987. Standard test method for moisture, ash, and organic matter of peat and other organic soils.

[34] ASTM C326: 1997. Test Method for Drying and Firing Shrinkage of Ceramic Whiteware Clays, American Society for Testing and Materials.

[35] ASTM C373:1994a. Test Method for Water Absorption, Bulk Density, Apparent Porosity, and Apparent Specific Gravity of Fired Whiteware Products. American Society for Testing and Materials.

[36] UNE-EN 772-11: 2011. Methods of test for masonry units – Part 11: Determination of water absorption of aggregate concrete, manufactured stone and natural stone masonry units due to capillary action and the initial rate of water absorption of clay masonry units.

[37] UNE EN 772-1: 2011. Methods of test for masonry units – Part 1: Determination of compressive strength.

[38] US Environmental Protection Agency, Method 13-11 Toxicity Characteristics Leaching Procedure (TCLP), 1992. Federal Register, Washington, DC, vol. 51, 11798-11877.

[39] Hongtao, H., Qinyan, Y., Yuan, S., Baoyu, G., Yue, G., Jingzhou, W., Hui, Y., 2012. Preparation and mechanism of the sintered bricks produced from Yellow River silt and red mud. *J. Hazard. Mater.*203-204, 53-61.

[40] Cabrera, M., Galvin, A.P., Agrela, F., Carvajal, M.D., Ayuso, J., 2014. Characterization and technical feasibility of using biomass bottom ash for civil infrastuctures. *Const. Build. Mater.* 58, 234-244.

[41] Suárez-Garcia, F., Martínez-Alonso, A., Fernández-Llorente, M., Tascón, J.M.D., 2002. Inorganic Matter characterization in vegetable biomass feedstocks. *Fuel* 81, 1161-9.

[42] Vassilev, S.V., Vassileva, C.G., Karayigit, A.I., Bulut, Y., Alastuey, A., Querol, X., 2005. Phase–mineral and chemical composition of composite

samples from feed coals, bottom ashes and fly ashes at the Soma power station, Turkey. *Turkey Int. J. Coal Geol.* 61, 35-63.

[43] Karayigit, A.I., Gayer, R.A., Querol, X., Onacak, T., 2000. Contents of major and trace elements in feed coals from Turkish coal-fired power plants. *Int. J. Coal Geol.* 44, 169-84.

[44] Querol, X., Fernandez-Turiel, J., Lopez-Soler, A., 1995. Trace elements in coal and their behaviour during combustion in a large power station. *Fuel* 74, 331-343.

[45] Haiying,Z.,Youcai, Z., Jingyu, Q., 2011. Utilization of municipal solid waste incineration (MSWI) fly ash in ceramic brick: Product characterization and environmental toxicity. *Waste Manage.* 31, 331-344.

[46] Demirbas, A., 2005. Potential applications of renewable energy sources, biomass combustion problems in boiler power systems and combustion related environmental issues. *Prog. Energy Combust.* 31, 171-192.

[47] Demir, I., 2008. Effect of organic residues addition on the technological properties of clay bricks. *Waste Manage.* 28, 622-627.

[48] Wagh, A.S., Poeppel, R.B., Singh, J.P., 1991. Open pore description of mechanical properties of ceramics. *J. Mater. Sci.* 26 3862-3868.

[49] Yakub, I., Du, J., Soboyejo, W.O., 2012. Mechanical properties, modeling and design of porous clay ceramics. *Mater. Sci. Eng. A.* 558, 21-29.

[50] Aouba, L., Bories, C., Coutand, M., Perrin, B., Lemercier, H., 2016. Properties of fired clay bricks with incorporated biomasses: Cases of Olive Stone Flour and Wheat Straw residues. *Constr. Build. Mater.* 102, 7-13.

[51] Kazmi, S.M.S., Abbas, S., Saleem, M.A., Munir, M.J., Khitab, A., 2016. Manufacturing of sustainable clay bricks: Utilization of waste sugarcane bagasse and rice husk ashes. *Constr. Build. Mater.* 120, 29-41.

[52] ASTM C62-10: 2010. Standard Specification for Building Brick (Solid Masonry Units Made From Clay or Shale), ASTM International, West Conshohocken, PA.

[53] Aouba, L.; Bories, C.; Coutand, M.; Perrin, B.; Lemercier, H., 2013. Properties of fired clay bricks with incorporated biomasses: cases of olive stone flour and wheat straw residues *Constr. Build. Mater.* 40, 390-396.

[54] ASTM C 67-03: 2003. Standards tests method for sampling and testing bricks and structural clay tile. American Sociaty for Testing and Material, PA, US.

[55] Pel, L.; Kopinga, K.; Bertram, G.; Lang, G., 1995. Water absorption in a fired-clay brick observed by NMR scanning. *J. Phys. D Appl. Phys.* 28, 675-680.

[56] RL-88. 2004. General specification for reception of the ceramic in the construction works.

[57] Ministerio de Agricultura, Alimentación y Medio Ambiente, Orden AAA/661/2013 sobre eliminación de residuos en vertederos [Ministry of Agriculture, Food and Environment, Order AAA/661/2013 on waste disposal in landfills]. MAM, BOE no. 97 23/4/2013. Spain. http://www.boe.es/diario_boe/txt.php?id=BOE-A-2013-4291.

Biographical Sketch

Dolores Eliche Quesada

Affiliation: Department of Chemical, Environmental, and Materials Engineering

Education: PhD in Science

Research and Professional Experience: I am Associate Professor of Materials Science and Metallurgical Engineering of the University of Jaen, Chemical Engineer by the University of Granada and PhD in Sciences by the University of Malaga. My research is mainly focused in the science and technology of materials, and in particular in the study and development of eco-efficient products through the valorization of industrial wastes for building sector (ceramic bricks). I am author and co-author of more than 35 papers in an international peer-reviewed journals. I have presented my research activity in a large number of national and international congresses. I act as a Reviewer for several peer reviewed journals like Construction and Building of Materials, Ceramic International, Applied Clay Science and Waste Management. I am Associate Editor of Journal of Minerals and Materials Characterization and Engineering.

Publications Last 3 Years

[1] D. Eliche Quesada, J. Leite-Costa, 2016. Use of bottom ash from olive pomace combustion in the production of eco-friendly fired clay bricks. *Waste Management* 48 (2016) 323-333. http://www.sciencedirect.com/science/article/pii/S0956053X15302245.

[2] S. Martínez-Martínez, L. Pérez-Villarejo, D. Eliche-Quesada, B. Carrasco Hurtado, P.J. Sánchez-Soto, G. N. Angelopoulus. Ceramics from clays and

by-product from biodiesel production: Processing, properties and microstructural characterization. *Applied Clay Science* 121-122 (2016) 119-126. http://www.sciencedirect.com/science/article/ pii/S0169131715301940.

[3] D. Eliche Quesada, R. Azevedo-da Cunha, F. A. Corpas-Iglesias. Effect of sludge from oil refining industry or sludge from pomace oil extraction industry addition to clay ceramics. *Applied Clay Science* 114 (2015) 202-211. http://www.sciencedirect.com/science/article/pii/S0169131715 002215.

[4] L. Pérez-Villarejo, S. Martínez-Martinez, B. Carrasco Hurtado, D. Eliche Quesada, C. Ureña Nieto, P. J. Sánchez-Soto. Valorization and inertization of galvanic sludge waste in clay bricks. *Applied Clay Science* 105 (2015) 89-99.http://www.sciencedirect.com/science/article/ pii/S0169131714005018.

[5] M.T. Cotes Palomino, C. Martínez García, F. J. Iglesias Godino, D. Eliche Quesada, F.A. Corpas Iglesias.Study of the wet pomace as an additive in ceramic material. *Desalination and Water Treatment* 54 (10) (2015) 1-7. http://www.tandfonline.com/doi/abs/10.1080/19443994.2015.1035678.

[6] M. T. Cotes-Palomino, C. Martínez-García, D. Eliche-Quesada, L. Pérez-Villarejo. Production of ceramic material using waste from brewing industry. *Key Engineering Materials.* 663(2015) 94-104. http://www.scientific.net/KEM.663.94.

[7] M. T. Cotes-Palomino, C. Martínez-García, F. J. Iglesias-Godino, D. Eliche-Quesada, F. J. Pérez La Torre, F. M. Calero de Hoces, F. A. Corpas-Iglesias. Study of waste from two-phase olive oil extraction as an additive in ceramic material. *Key Engineering Materials* 663 (2015) 86-93. Avalaible from: http://www.scientific.net/KEM.663.86.

[8] D. Eliche Quesada. Reusing of Oil Industry Waste as Secondary Material in Clay Bricks. *Journal of Mineral, Metal and Material Engineering* 1 (2015) 29-39. http://www.synchropublisher.com/jms/index.php/JMMME/ article/view/40.

[9] D. Eliche-Quesada, Francisco A. Corpas-Iglesias. Valorization of Liquid Effluents from Olive Oil Extraction Activity in the Production of Ceramic Bricks: Influence of Conformation Process. In: "*Fruit and Pomace Extracts: Biological Activity, Potential Applications and Beneficial Health Effects.*" Jason P. Owen Editor. New York, Nova Science Publisher Inc, pp. 29-52. 2015. ISBN 978-1-63482-510-8 (eBook). Available form: https://www.novapublishers.com/catalog/ product_info.php?products_id=54356&osCsid=.

[10] L. Pérez Villarejo, S. Martínez Martínez, D. Eliche Quesada, B. Carrasco Hurtado, P.J. Sánchez Soto. Mechanochemical Treatment of Clay Minerals

by dry Grinding: Nanostructured materials with enhanced surface properties and reactivity. In: "*Clays and Clay Minerals: Geological Origin, Mechanical Properties and Industrial Applications.*" 1-3, pp. 67-114. 2014. https://www.novapublishers.com/catalog/product _info.php?products_id=49439.

[11] D. Eliche-Quesada, F.J. Iglesias-Godino, L. Pérez-Villarejo, Francisco A. Corpas-Iglesias. Replacement of the mixing fresh water by wastewater olive oil extraction in the extrusion of ceramic bricks. *Construction and Building Materials* 68 (2014) 659-666. Available form: http://www.sciencedirect.com/science/article/pii/S095006181400 7326.

[12] D. Eliche-Quesada, Francisco A. Corpas-Iglesias.Utilisation of spent filtration earth or spent bleaching earth from the oil refinery industry in clay products. *Ceramics International*. 40 (2014) 16677-16687. http://www.sciencedirect.com/science/article/pii/S027288421401267X.

In: Ceramic Materials
Editor: Jacqueline Perez

ISBN: 978-1-63485-965-3

Chapter 2

NANOSTRUCTURED CERAMICS FOR THE CONTROL OF HEAT FLOWS

Evgenii I. Salamatov*[1]*, Oksana V. Karban*[1,*]*,
***Efim N. Khazanov*[2] *and Andrey V. Taranov*[2]**

[1]Physical-Technical Institute UB RAS, Izhevsk, Russia
[2]Kotelnikov Institute of Radio Engineering and Electronics RAS, Moscow, Russia

Abstract

Nowadays considerable attention is paid to the creation of a new type of nanostructured materials in which one can control the heat flow. Since it is believed that the basic element of such thermocrystals should be phononic lattices with a wide forbidden gap, it is an urgent task to find new methods for their synthesis that strikes a balance between the thermocrystal efficiency and ease of fabrication.

It is shown both theoretically and experimentally that the compacted ceramics can exhibit the properties of a phononic lattice, i.e., a forbidden gap may arise in the phonon spectrum. The position and width of the gap in such systems are determined by the average grain size of ceramics, as well as by the thickness and elastic properties of the grain boundaries. Inclusions of metal phase in dielectric matrix can create photonic traps for the nanocomposite materials and determine their diffusion ratio.

The approach for the synthesis of nanocomposites by use of ceramics technology was introduced. It allows to create materials with required

* E-mail address: ocsa123@yahoo.com (Corresponding author).

characteristics by the compaction and sintering process and to give a practical advice on synthesis of nanostructured ceramics with specified parameters.

Keywords: thermocrystals, heat flow, phononic lattice, metal-oxide composites, phonon spectroscopy

Introduction

The present day interest in heat transport by phonons in nanostructured mesoscopic systems is determined by the general tendency of electro-acoustic devices to shrink in size due to development of advanced technologies which allow one to prepare samples with almost atomic precision and to measure the thermal conduction of mesosystems comparable in size to the dominant phonon wavelength at temperatures of about 1 K (~100 nm). Temperature regime is of prime importance in operation of such devices. In this connection particular attention is now paid to the synthesis and study of thermocrystals, i.e., nanostructured materials in which the heat flow can be controlled [1, 2]. Since it is believed that phonon lattices with a wide band of forbidden frequencies in the spectrum region above 10^{11} Hz may form the basis of thermocrystals, the problem of creation of such systems with an acceptable price-quality relationship is very urgent.

One of the methods for obtaining phononic band gap structures is the compaction of ceramics from nanopowders. For the first time, the possibility of existence of a forbidden gap in the phonon spectrum of compacted oxide nanoceramics was suggested in [3]. This idea was then experimentally confirmed [4] in a study of the transport properties of weakly nonequilibrium phonons at helium temperatures by the phonon spectroscopy method [5]. It turned out that the anomalous growth with temperature of the phonon diffusion coefficient observed in ceramics with an average grain size less than 100 nm can be explained by the existence of such a gap. Use of this idea made possible to interpret the experimental evidence obtained when studying the structure and elastic properties of the interface boundaries of compacted ceramics by phonon spectroscopy [6-8]. All experimental results were obtained for standard samples of structural ceramics prepared by magnetic pulse compaction [9] or dry pressing [10]. This type of ceramics always exhibit a variation in grain size, but the theoretical studies performed [11, 12] have shown that the gap in the phonon spectrum is retained to sufficiently large values of the grain size dispersion ($\sigma \sim 0.5$), which can be implemented with modern technologies of nanoceramic synthesis. This is allows us to offer a new way of manufacturing phononic band gap structures by compacting [13].

Another way to control heat flow is creation of "traps" for heat phonons in the system. Such "traps" can be represented by separate atoms or their clusters, whose properties differ from matrix behaviour. In crystals these "traps" can be referred to defects which normally result in redundant low-temperature thermal capacity. For example, the "traps" may consist of magnetic atoms in non-magnetic matrix [14, 15] and defects of structure caused by various atom surrounding of vacancies in solid electrolytes [8]. In dielectric ceramics, these "traps" there may consist of nanosize spots of metal phase [7], whose electron specific heat is much higher at low temperatures than specific heat of the matrix. Application of the unique method of phonon spectroscopy for determination of transport characteristics of phonons allows to study both phonon spectrum of ceramics systems and effects connected to the presence of "traps." Research of ceramics structure and chemical-physical properties together with its synthesis makes it possible to investigate deeply the processes which cause the observed results.

In the present work we have summarized the research done by our team in this field.

Phonon Propogation in Nanoceramics. Theoretical Model

Ceramic materials with nanograins have a complex of mechanical and functional properties, making them promising for a wide range of practical applications. An effective method for studying such systems is phonon spectroscopy (thermal-pulse method), which is based on the analysis of the time dependence of propagation of a pulse of weakly nonequilibrium phonons at liquid-helium temperatures [16]. At such temperatures, the wavelengths of injected phonons are in the range 10–50 nm; i.e., they are comparable with the sizes of different fragments of the ceramic structure (grains, pores, grain boundaries). For adequate interpretation of experimental data, the development of a theoretical model describing the transport of nonequilibrium phonons in nanostructured ceramic materials is required. However, this is a difficult problem due to the absence of translational symmetry in such systems. A possibility of studying the structure of grain boundaries in a dense dielectric ceramics was shown in [16] for the model of an isotropic material, in which scattering of a phonon passing through a planar boundary between two layers was calculated using the method of acoustic matching. The assumption of a planar boundary (the radius of curvature R tends toward infinity) indicates that this model can be applied only to coarsegrain ceramics. Indeed, within this model, one cannot describe the experimentally observed falloff of the linear dependence of the phonon diffusion coefficient on the grain size, $D(R_g)$ which occurs when the average grain size of the ceramic becomes comparable with the wavelength of injected phonons [6]. To take into account the finite boundary curvature, spherical shells randomly distributed in space and having elastic properties differing from those of grains were considered in [3, 6].

As theoretical model we considered an elastic medium with a density ρ_0, whose elastic properties are described by one modulus of elasticity K_0 (scalar model) with the dispersion relation $\omega(q) = v_0 q$ ($v^2_0 = K_0/\rho_0$) for phonons of all polarizations. We will consider spherical shells with an external radius R_g, thickness d ($d/R_g << 1$), and the elasticity parameters v_1 and ρ_1 ($K_1 = v^2_1 \rho_1$), as main scattering centers, which model grain boundaries. The material within a shell and beyond it is assumed to be that of ceramic grains. In the case of multiphase ceramics, the elasticity parameters of the material within the shell (v_2, ρ_2)

$$G(q,\omega) = \frac{1}{\omega^2 - \omega^2(q) - \Sigma(\omega,q)} \tag{1}$$

where $\Sigma(\omega,q)$ is the energetic part, which exists due to elastic scattering of phonons from inhomogeneities. In this study, we will restrict ourselves to the linear (with respect to the concentration of scattering centers) approximation, where $\Sigma(\omega, q) = ct_g(\omega)$, where c is the relative concentration of scattering centers and $t_g(\omega)$ is the diagonal element of the one-center scattering matrix. To determine $t_g(x)$, we solved the standard problem of scattering of a plane wave on spherical inclusion. The calculations gave the expression:

$$t_g(x) = \frac{6xv_0^2}{R_g^2} \sum (2l+1) a_l \,, \tag{2}$$

$$a_l = \frac{m_1^2 \mu_2 A_l j_{l_0} - m_1 \mu_1 (m_2 j_{l_0} B_l + \mu_2 j'_{l_0} C_l) + \mu_1^2 m_2 j'_{l_0} D_l}{m_1^2 \mu_2 A_l h_{l_0} - m_1 \mu_1 (m_2 h_{l_0} B_l + \mu_2 h'_{l_0} C_l) + \mu_1^2 m_2 h'_{l_0} D_l} \,, \tag{3}$$

$$A_l = j_{l3}(h'_{l1} j'_{l2} - j'_{l1} h'_{l2}) \,,\ B_l = j'_{l3}(h'_{l1} j_{l2} - j'_{l1} h_{l2}) \,,\ C_l = j_{l3}(h_{l1} j'_{l2} - j_{l1} h'_{l2}) \,,$$

$$D_l = j'_{l3}(h_{l1} j_{l2} - j_{l1} h_{l2})$$

Here, $x = qR_g$; $m_s = v_0/v_s$; $\mu_s = \rho_s/\rho_0$ $(s = 1, 2)$; and j_{li}, h_{li}, j'_{li}, and hj'_{li} are the spherical Bessel and Hankel functions of order l and their derivatives taken with the arguments $i = 0 - qR_g$, $i = 1 - m_1qR_g$, $i = 2 - m_1q(R_g - d)$, and $i = 3 - m_2q(R_g - d)$. Expressions (2) and (3) have a general form and, in the limitation $d = 0$, is applicable for spherical inclusions.

For the diffusion coefficient of phonons described by the Green's function (1), we have [17]

$$D(x,R_g) = \frac{2v_0^3}{R_g\Phi}\frac{x}{c\mathrm{Im}t(x)}, \tag{4}$$

where c is the relative concentration of scattering centers and $\Phi = \partial\ln(x^2 + c\mathrm{Re}tg(x))/\partial\ln x^2$. The appearance of this term in the expression for the diffusion coefficient is related to the renormalization of the dispersion relation for phonons in the presence of scattering centers. The dispersion relation for phonons ω^* propagating in an inhomogeneous medium is found from the following condition

$$\omega^2(q) - \omega^{*2} - c\mathrm{Re}t(\omega^{*2}) = 0, \tag{5}$$

It was shown in [17, 18] that resonance scattering of long-wavelength phonons leads to a significant reconstruction of the phonon spectrum and, at a sufficiently high defect concentration, a low-frequency gap arises in the dispersion relation for acoustic phonons. To reveal the conditions for resonant scattering from spherical shells, we analytically investigated the expression for the zero harmonic from (2), which makes the main contribution to the scattering matrix at low frequencies. It is shown that resonant scattering from spherical shells may occur when the condition $K_1/K_0 \ll 1$ is satisfied. Leaving only the zero-harmonic contribution to the scattering matrix for the dimensionless scattering cross section

$$\sigma^*(x) = \frac{\sigma(x)}{2\pi R_g^2} = \mathrm{Im}\frac{t_g(x)R_g^2}{3xv_0^2}, \tag{6}$$

we obtain

$$\sigma^* = \frac{2x + (\Gamma x^3)}{x^2 - x_r^2 - i\Gamma x^3}, \tag{7}$$

where

$$x_r = \sqrt{K_1 R_g / K_0 d}\,, \tag{7a}$$

is the resonance frequency, Γ is the resonance width, and the term in parentheses in the numerator in (7) provides “matching” between resonant and geometric scattering (σ* = 1) at large *x*.

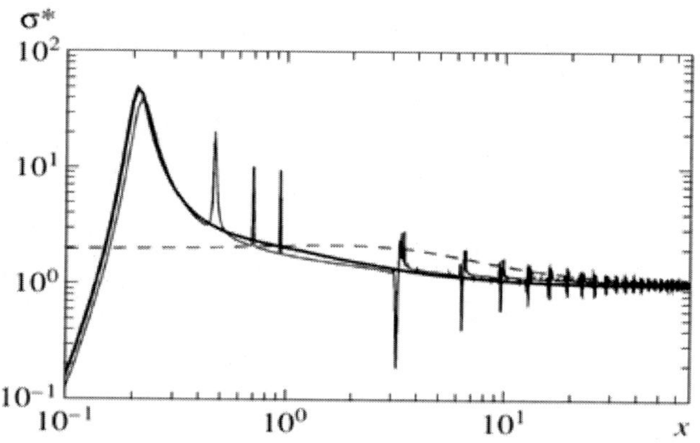

Figure 1. Normalized scattering cross section at $d = 0.2$, $v_l = 0.2$, and $\rho_l = 0.2$ ($x_r \approx 0.2$), calculated using expressions (2) and (3) (thin line), and its approximation (7) (bold line). The dashed line corresponds to the scattering cross section for pores with $R_p = 0.25$ and $c_p = 0.1$.

The dimensionless scattering cross section, calculated from formula (7) with the use of expressions (2) and (3) is shown in Figure 1. The figure indicates that approximation (7), retaining the general form of the dependence $\sigma^*(x)$, does not take into account the resonances of higher order, which, in case $x > 1$, degenerate into the resonances corresponding to phonon scattering from planar grain boundaries at frequencies $x_l = l\pi v_1 R_g / v_0 d$, which reflects the essence of the model [4] at normal angles of incidence of nonequilibrium phonons. To obtain useful analytical expressions, in this section we will apply approximation (7) and dimensionless units, assuming that $R_g = v_0 = \rho_0 = 1$. In addition, we will consider the case of a single-phase ceramic: $v_0/v_2 = \rho_2/\rho_0 = 1$.

Analysis of the phonon spectrum of disordered systems at resonant phonon scattering from defects was performed in detail in [17,18]. It was shown that, at a sufficiently high defect concentration, cross splitting of the acoustic phonon branch is observed in this case, and the phonon spectrum exhibits a gap, i.e., a frequency range in which phonons cannot propagate. According to [17], the bottom and top edges of this gap are determined, respectively, by the expressions

$$\omega_{bot} \approx \omega_r$$

$$\omega_{top} \approx \sqrt{\omega_r^2 + A(n)}$$

where ω_r is the resonant scattering frequency. $A(n)$ depends on the concentration of scattering centers and is determined by the lattice parameter of the effective ideal crystal. In other words, the phonon spectrum of a disordered system is similar to the spectrum of a phononic crystal with a periodicity parameter determined by the concentration of spherical scattering centers of different nature and the corresponding frequencies of Bragg reflections, near which gaps can arise.

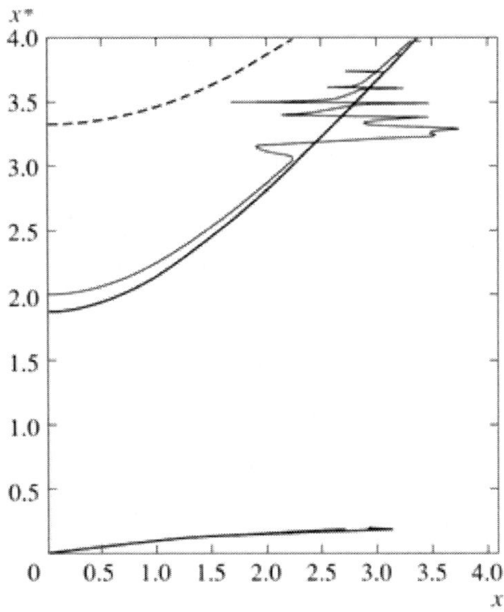

Figure 2. Phonon dispersion relation at the same values of the parameters as in Figure 1.

In our case, the renormalized dispersion relation for phonons in the limit $\Gamma \to 0$ has the form [17]

$$x^{*2} = x^2 + x_r^2 + 6c_g \pm \sqrt{(x^2 + x_r^2 + 6c_g)^2 - 4x_r^2 x^2}\,, \tag{8}$$

where c_g is the volume fraction of spheres with a radius R_g. Hence, we have the following expressions for the gap edges:

$$x_{bot,g} \approx x_r \;\; x_{top,g} \approx \sqrt{x_r^2 + 6c_g}\,, \tag{9}$$

The phonon dispersion relation, calculated from formula (8) with the use of expressions (2) and (3), is shown in Figure 2.

After the simplifications made, one can easily perform formal analysis of the frequency behavior of the diffusion coefficient with respect to the position of two remaining: resonant level x_r and the defect concentration c_g. The parameter x_r is determined by the size and elastic properties of shells. According to expression (7a), resonant scattering from shells may occur only in poor-quality ceramics with thick loose boundaries, and one should expect absence of resonant scattering at high-temperature sintering (which stabilizes boundaries). Here, c_g determines the volume fraction of grains with resonant shells and is a parameter of the model. Figures 3 and 4 show the dependences $D(x)$ at different values of these parameters. For all curves, Rayleigh scattering ($D \sim 1/q^4$) is observed at small x; at large x, geometric scattering occurs:

$$D = l_{tr} v_0 \;\; l_{tr} = 2R_g / 3c_g\,, \tag{10}$$

Note that asymptotics (10) is also valid for nonresonant shells. In the intermediate region ($x \sim 1$), at small x_r (large c_g), a gap arises, whose edges are determined by (9); this gap becomes a minimum with an increase in x_r (decrease in c_g). The upper curve in Figure 3, calculated using expressions (2) and (3), is related to a system containing absolutely solid spheres (well-stabilized boundaries). In this case, a gap is absent at the transition from the Rayleigh to geometric scattering.

Figures 3 and 4 indicate that, in the range of x close to unity, the behavior of the phonon diffusion coefficient may exhibit qualitative changes, including object opacity for phonons and a change in the sign of the derivative $\partial D/\partial T$, with a change in the elasticity parameters of the boundary layer of the ceramics (which determine x_r) or the concentration of scattering centers.

Along with grain boundaries, pores are also effective scattering centers in ceramics. There is an analog of such a situation in the lattice model: a substitutional impurity atom with an infinitely large mass [17]. Within the approximation used, the scattering matrix formally coincides with the resonant matrix (7) at $x_r = 0$. In this case, the coordinates of the bottom and top edges of the gap are, respectively, zero and $x_{top,p} = \sqrt{6c_p} R_g / R_p$ (c_p is the volume fraction of pores and R_p is their radius). The dimensionless scattering cross section and the dispersion relation for phonons for a system with pores are shown in Figures 1 and 2 by dashed lines.

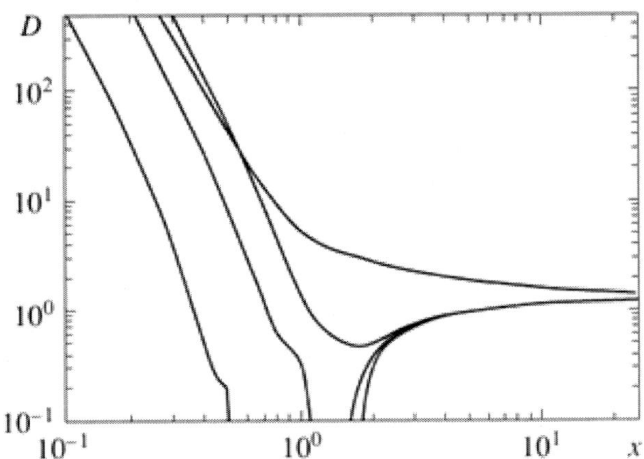

Figure 3. Diffusion coefficient *D(x)* calculated at c_g = 0.5 for different resonant frequencies. x_r = 0.5 for different resonant frequencies. From bottom to top: x_r = 0.5, 1.0, and 1.5; the upper curve corresponds to the case of absolutely solid spheres.

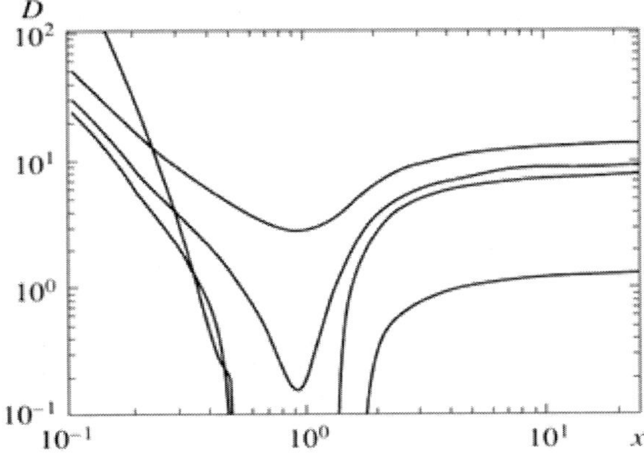

Figure 4. Diffusion coefficient *D(x)* calculated at x_r = 0.5 for different resonant frequencies. From bottom to top: c_g = 0.20.35, 0.4, and 0.5.

Let us consider in more detail the case where phonon scattering occurs from both shells and pores, remaining within the linear approximation with respect to concentration:

$$\Sigma = c_g t_g + c_p t_p$$

In the mode of geometric scattering (x >> 1), where the factor Φ from (4) tends to unity, the transport length satisfies Matthiessen's rule

$$l_{tr} = \frac{l_p l_g}{l_p + l_g} = \frac{2}{3} \frac{R_g R_p}{c_g R_p + c_p R_g}, \tag{11}$$

In the range ($x \sim 1$), interference corrections arise because Φ is also additive in concentration and the gap top edge extends from zero to the frequency

$$x_{top}^2 = \tfrac{1}{2}(x_{top,g}^2 + x_{top,p}^2 + (\sqrt{(x_{top,g}^2 + x_{top,p}^2)^2 - 4x_r^2 x_{top,p}^2})^{1/2})$$

Obviously, similar results will be obtained in the presence of other scattering centers. A factor of fundamental importance is that inclusions with a characteristic size smaller than the grain size ($R_g / R_p > 1$) can play an important role even at low concentrations, shifting the gap (or minimum) to higher frequencies. Formation of a gap in the phonon spectrum of a material in the presence of several types of scattering centers is considered below in more detail (Figure 10) by a specific example.

Method of Studying the Kinetic Characteristics of Thermal Phonons at Liquid-Helium Temperatures

The experimental technique and the analysis of propagation of nonequilibrium phonons in ceramic materials were reported in [16]. In this study, the samples were plane-parallel polished plates with a thickness of 0.1–1.0 mm and an area of about 0.5 cm^2. Phonon injector (gold) and detector (tin) films were formed on the opposite faces of the samples by thermal deposition in vacuum. Experiments were performed in liquid helium in the temperature range 1.5–3.8 K. The temperature was controlled by pumping off helium vapor. The temperature measurement accuracy was no worse than 10^{-3} K. The operating point of the bolometer was shifted by applying a weak magnetic field. The experimentally measured value was the time tm of arrival of the maximum diffusion signal of nonequilibrium phonons at the detector (bolometer). Phonons were injected from the metal film located on the opposite face of the sample and heated by a short (10^{-7} s) current pulse to a temperature T_h; $\Delta T = T_h - T_0 << T_0$ (T_0 is the thermostat temperature). This approach made possible to obtain the temperature dependence $t_m(T)$ to change the thermostat temperature, assuming that $T_h \approx T_0$.

For the "planar" source geometry, the diffusion character of the propagation of nonequilibrium phonons (NPs) is described by the equation

$$\frac{\partial^2 T}{\partial x^2} = \frac{1}{D}\frac{\partial T}{\partial t}$$

where $D = \kappa/c_v$ is the diffusion coefficient of phonons and κ and c_v are the thermal conductivity and specific heat of the material, respectively. In experiments, we measured the time t_m of arrival of the maximum diffusion signal at the bolometer, $t_m = L^2/2D$, where $D = lv/3$, l is the mean free path of NPs, v is the phonon velocity averaged over polarizations, and L is the sample size in the direction of NPs propagation.

Resonant Scattering of Nonequilibrium Phonons in Nanostructured Ceramics Based on YSZ + Al_2O_3 Composites

In the work [6] the transporting behavior of ceramics Al_2O_3, the solid solution ZrO_2:Y_2O_3 (YSZ) and their composites was studied. Figure 5 shows the dependences $D(R)$ at $T = 3.8$ K for single-phase ceramics: Al_2O_3 and the solid solution ZrO_2:Y_2O_3 (YSZ). In both cases, sharp falloff in the $D(R)$ dependence is observed at $qR \sim 20$ (for Al_2O_3, $v = 7.4 \times 10^5$ cm/s; for YSZ, $v = 4.33 \times 10^5$ cm/s), which indicates the onset of effective scattering of nonequilibrium phonons by ceramic grains at R ~ 100 nm.

According to the above-described model, the presence of a gap in the phonon spectrum suggests that the parameter qR ~ 1 can be used; i.e., at $q \sim 10^6$ cm^{-1}, the grain size should be 10–20 nm.

The current absence of such single-phase nanostructured ceramics based on the noted oxides does not exclude the possibility of composites formation on their basis, which, along with the main fraction of stable grains (crystallites), contain a certain amount of the nanostructured phase of another material. Such composites (YSZ + Al_2O_3) were synthesized in [18].

The feature of this structure is the presence of a finegrained metastable phase of corundum (R = 20–40 nm), which is partially transformed into a denser α-Al_2O_3 phase (ρ = 3.97 g/cm^3) with an increase ofthe synthesis temperature T_s. This transition leads to the formation of nanoscale shrinkage pores, whose number is related to the content of the metastable Al_2O_3 phase. The presence of additional nanoscale scattering centers (metastable corundum phase, shrinkage pores), even with relatively low concentrations, may lead, as was shown above, to a significant reconstruction of the phonon spectrum—formation of a gap and shift of its top edge to higher frequencies.

Along with the 9.8% YSZ + 14.3% Al_2O_3 composites, we investigated the 9.8% YSZ single-phase ceramic (samples 712, 713, 720, and 721), synthesized at the same temperatures T_s. The parameters of the samples studied are listed in Table 1, which contains also the parameters for YSZ samples 236 and 798 from another series, stabilized with 2.8 and 4.1% Y_2O_3 but with a smaller grain size R_g. The pressing technique, firing technology, and study of the microstructure and phase composition of the samples were described in detail in [18].

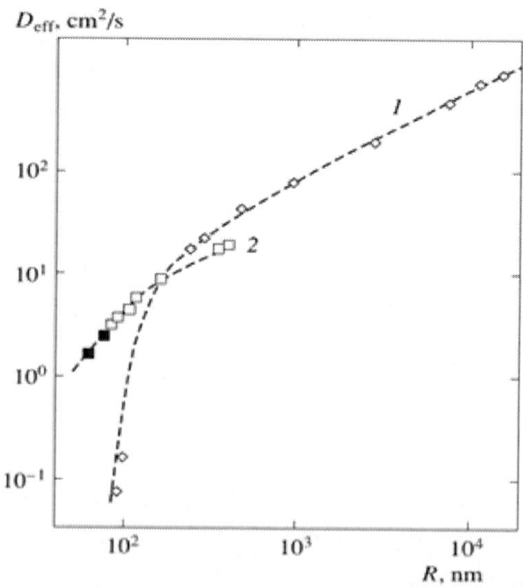

Figure 5. Dependence of the effective diffusion coefficient of phonons on the average grain size at T = 3.8 K for (1) Al_2O_3 and (2) YSZ ceramic samples. Black squares correspond to YSZ samples 236 and 798.

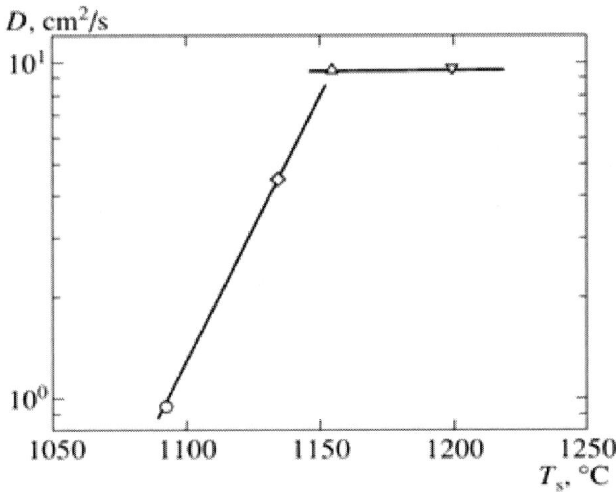

Figure 6. Dependence $D(T_s)$ at T = 3.8 K for 9.8% YSZ. Sample nos. (see Table 1): 721 (o), 713 (◊), 712 (Δ), 720(∇).

Figure 6 shows the dependence of $D(T_s)$ at T = 3.8 K for 9.8% YSZ samples. The absence of nanostructured fragments in these samples (except for grain boundaries); a significant average grain size (R_g > 100 nm), corresponding to the linear portion of the dependence $D(R)$ in Figure 5 (except for sample 721, which was synthesized at T_s = 1365 K); and the value of the ratio $l_{tr}/R_g \geq 1$ indicates that scattering of nonequilibrium phonons is mainly determined by the structure of grain boundaries, which are stabilized with an increase of T_s .

Table 1.

Sample no	**T_s, K**	**ρ,g/cm^3**	***R*,nm YSZ**	**L, 10^{-2}cm**	**Content of Al_2O_3 phases**	
					α phase, wt%	**(δ+γ) phases,wt%**
721	1365	5.49	84	2.65	0	0
713	1413	5.74	110	3.2	0	0
775	1413	5.22	92	2.0	25	75
712	1428	5.75	135	2.1	0	0
769	1428	5.21	64	2.1	32	68
720	1473	5.84	132	3.3	0	0
776	1473	5.26	84	2.75	62	38
236	1373	6.1	63	4.2	0	0
798	1423	6.1	78	3.8	0	0

In addition, it can be seen in Figure 6, that at $T_s \geq 1428$ K, boundaries between grains of the YSZ phase are well stabilized; i.e., nonequilibrium phonon scattering from boundaries is independent of T_s.

Figure 7 shows the dependence of measured temperature vs. arrival time of the maximum signal of nonequilibrium phonons, normalized to L^2, in the samples of 9.8% YSZ + 14.3% Al_2O_3 composites and their satellites—samples of cubic oxide 9.8% YSZ at three values of T_s. The character of the dependence $t_m(T)$ in the composites differs both in magnitude and sign $(\partial t_m/\partial T)$, even when boundaries between grains of the primary phase YSZ are already sufficiently well stabilized (samples 769, 776). For all single-phase YSZ ceramic samples, $\partial t_m/\partial T > 0$.

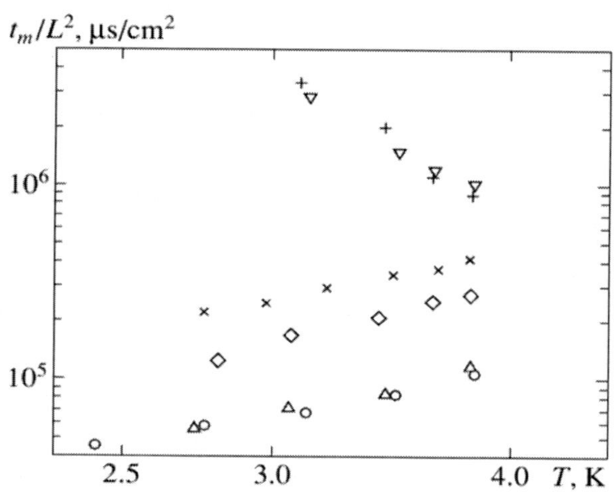

Figure 7. Temperature dependences of the ratio t_m/L^2. Sample nos. (see Table 1): 712 (o),720 (Δ), 769 (∇), 776 (×),775 (+), 713 (o).

Table 2.

Sample no	T_s, K	c_g	R_g ,nm	v_g/v_b	ρ_g/ρ_b	d, nm
712	1428	0.74	135	1.5	2	2.5
720	1473	0.74	132	2	2	1
236	1373	0.74	63	2	1.75	6.5
798	1425	0.5	78	2	1.75	4.5

Let us first compare the temperature dependences of the diffusion coefficient of nonequilibrium phonons $D(T) \sim 1/t_m$ in single-phase YSZ ceramic samples within the predictions of the theory considered above (Figure 8). Along with samples 712 and 720, characterized by similar grain sizes and the character of boundary stabilization, for which the conditions $qR \gg 1$ and $l_{tr}/R_g \gg 1$ are satisfied (i.e., scattering of nonequilibrium phonons is determined mainly by grain boundaries [4]), we will consider single-phase 236 and 798 samples with smaller values of R_g (Figure 5, curve 2, black squares). The experimental dependences for these samples with $\partial t_m/\partial T < 0$ corresponds to the theoretical behavior on the right from the minimum (Figures 3, 4). In this case, the increase in the grain size at higher T_s (sample 798) is insufficient to explain the increase in the diffusion coefficient by a factor of 2 in comparison with sample 236. It is natural to suggest (in accordance with the results of the calculation using expressions (2) and (3)) that the character of scattering of nonequilibrium phonons is determined not only by the value of R_g but also by the decrease in the concentration of resonance scattering centers (c_g = 0.5). The obtained data on the thickness of grain boundaries (Table 2) differ but do not contradict the estimates that were made in [4] for the single-phase ceramics YSZ on the basis of the model [16], which does not take into account the finite curvature of actual grains.

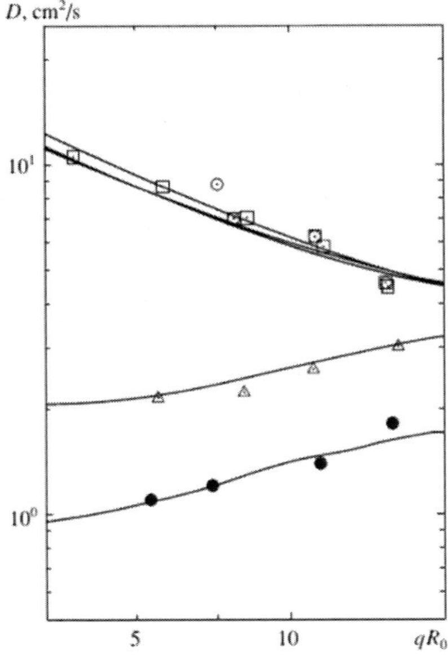

Figure 8. Diffusion coefficient $D(x)$ of single-phase ceramics (R_0 = 100 nm). Sample numbers. (see Table 1): 712 (□), 720 (ʘ), 236 (•), 798 (Δ).

The dependence $D(T)$ (Figure 9) for the 9.8% YSZ +14.3% Al_2O_3 composites differs both in magnitude and behavior from the corresponding dependence for the YSZ samples. The reason is that these composites, along with VSZ matrix grains, contain other nanoscale defects: grains of the metastable Al_2O_3 phase and pores. It was shown in [9] that the growth of particles of metastable Al_2O_3 phases in a cubic oxide matrix is hindered, and, at $T_s < 1473$ K, the size of metastable phase particles does not exceed 20–40 nm. In the same time, in the temperature range T_s = 1373–1473 K, the α-Al_2O_3 phase grows most intensively. This process is accompanied by shrinkage with formation of nanoscale shrinkage pores, which grow with an increase in T_s. The formed α-phase grains had the size R = 80 nm at T_s = 1473 K. Thus, the pattern of formation of the structure of 9.8% YSZ + 14.3% Al 2O3 composites, synthesized at T_s = 1413–1473 K, can be represented as follows. A matrix composed of grains (crystallites) of the cubic YSZ phase with $R \geq 100$ nm contains nanoscale fragments of metastable Al_2O_3 phases, which, with an increase in T, are partially transformed into the α-Al_2O_3 phase, undergoing shrinkage. As a result, nanoscale shrinkage pores are formed, which are localized near Al_2O_3 grains and increase in size with an increase in T_s.

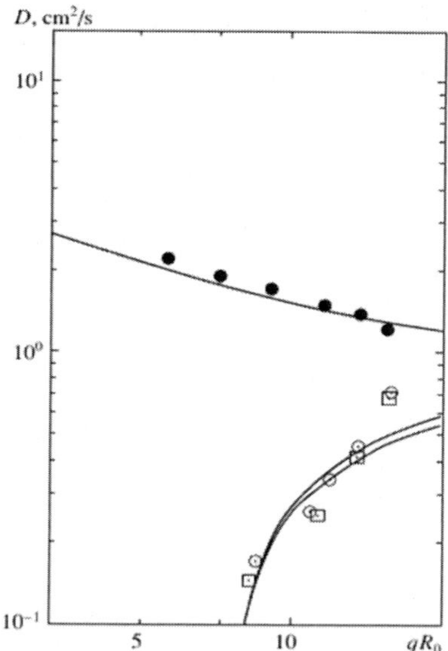

Figure 9. Diffusion coefficient *D(x)* of composites (R_0 =100 nm). Sample numbers (see Table 1): 775 (□), 769 (ʘ), 776 (•).

An increase in the annealing temperature to T_s =1473 K leads to a decrease in the concentration of nanostructured defects, which actively scatter nonequilibrium phonons. Therefore, the dependence *D(T)* with a sharp rise (derivative $\partial D/\partial T > 0$, see Figure 9), which corresponds to the top of the gap in the phonon spectrum in samples 775 and 789, is transformed for sample 776 (synthesized at 1473 K) into a dependence with $\partial D/\partial T < 0$, and scattering of nonequilibrium phonons in this sample begins to be controlled predominantly by grain boundaries, as in samples 712 and 720. The presence of nanoscale inclusions in multiphase ceramics leads to the formation of a gap in the phonon spectrum, which is due to the resonant scattering of phonons from these defects. The formation of a gap in the phonon spectrum of samples of 775 and 769 is clarified in Figure 10, which shows (for specific parameters of sample 775) the diffusion coefficients due to individual defects: pores ($c_g = c_i = 0$, $c_p = 0.05$), shells ($c_g = 0.74$, $c_i = c_p = 0$), and shells around metastable Al_2O_3 phases ($c_g = c_p = 0$, $c_i = 0.21$), as well as the total diffusion coefficient for sample 775.

Table 3.

Sample no.	T_s, **K**	R_g, **nm**	R_p, **nm**	R_i, **nm**	v_g/v_b (ρ_g/ρ_b)	d_b, **nm**	v_g/v_{bi} (ρ_g/ρ_{bi})	d_{bi}, **nm**
775	1413	92	11.5	20	2.5	8	3	6
769	1428	64	15	20	2.5	8	3	5
776	1473	84	50	45	2.25	4	2.25	4

Note: v_g (ρ_g) is the velocity (density) in a YSZ grain; the subscripts *b* and *bi* indicate boundaries around YSZ and corundum grains, respectively.

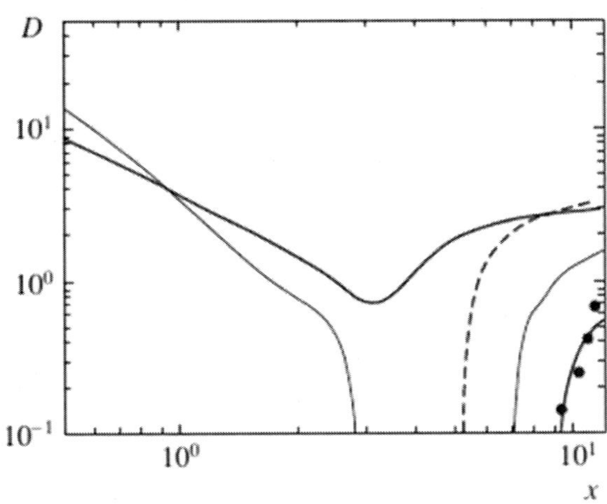

Figure 10. Diffusion coefficients related to pores (dashed line), shells (semibold line), and shells around Al_2O_3 grains (thin line) and the total diffusion coefficient (bold line) for sample 775.

Evolution of the parameters of the composite ceramic with an increase in the sintering temperature is shown in Table 3. As follows from this table, a fairly large number of parameters is necessary to compare the experimental dependences *D(T)* for samples investigated here with the theory predictions. Therefore, it was assumed that the volume fraction of different phases did not change from sample to sample: $c_g = 0.74$, $c_i = 0.21$, and $c_p = 0.05$. The radii of matrix grains (R_g), Al_2O_3 grains (R_i), and pores (R_p) were determined from independent experiments [18]. Other parameters were found by fitting the calculated curves to the experimental dependences. In the case of presence of a gap or minimum in the curve *D(x)* (samples 775, 769, 236, and 798), the main problem parameter is the value x_r from expression (7a), which is determined by the shell characteristics ρ_1, v_1, and *d*.

Kinetics of Thermal Phonons and the Structure of Nanodispersed Iron-Containing Corundum-Based Cermets

The presence of a metallic fraction in nanocomposites can substantially change the transport properties of subterahertz phonons due to both resonance scattering by structural fragments and an effective electron–phonon interaction in metallic grains at low temperatures. The purpose of this section is to study the transport properties of subterahertz phonons and the phonon spectrum and structure of Al_2O_3-based iron-containing nanocermets.

The main problem during synthesis of nanostructured composite materials based on oxides and metal phase is obtaining high-disperse powders with homogeneous distribution of metal phase and nanosized grains of both metal and ceramic phase, and also preserving this homogeneous nanosized structure in sintered bulk composite material. Moreover, due to the difference of thermal expansion ratio of metal and ceramics, it is necessary to use ceramic particles of extremely small size to obtain composites of high thermal stability.

The method of mechanical activation [20, 21] is used to increase dispersibility of initial powders and their mixes. Milling of powders using different types of high-energy mills, including attritors, vibration mills and planetary ball mills, may result in dramatic change of chemical and physical properties of powders, in particular in decreased period of sintering which subsequently leads to slow growth of grains but only at short periods of treatment. In case of long treatment of powders, their sintering gets worse, at this non-repeatability of ceramics density occurs which is connected with formation and condensation of aggregates during the process of mechanical activation. Choice of the dispergation method is determined by the required properties and structure of material [20, 21].

It is also a serious problem to compact dense pressed nanosized powders with regular volume density, as nanopowders are difficult to press and traditional methods of static pressing do not provide density high enough. Physically, poor pressing ability of nanopowders is caused by adhesion forces between particles, relative value of which dramatically increases with less particle size. Consequently the component of inter-particle friction in a pressed powder body is increased significantly. At the same time dust-like nanopowders are characterised by low apparent density due to large air volume that screens the surface of nanoparticles. Besides nanoceramic powders usually contain solid agglomerates (strongly connected particles) which need to be destructed before redistribution during the process of pressing. Thus, in order to preserve nanostructure while obtaining high enough density, it is necessary to apply special high-energy methods of consolidation to suppress intense recrystallisation and possible residual pores.

Also the temperature and sintering time as well as heating method and gaseous atmosphere content can have an influence on the grain size, phase content, structure and composition of interface areas of composites and finally on the properties of the material being synthesized.

That is why it was used the new approach to obtain composite thermocrystals made of nanosized aluminium oxide powders with iron addition (up to 25 vol.%). This method provides bulk cermets, fine-grained and homogeneous. It consists of two points.

1. As opposed to standard schemes of preparation of composites of aluminium and iron oxide out of aluminium, aluminium oxide and iron oxide powders by means of iron reduction during mechanical synthesis [20-23], there was performed mechanical activation of initial mix of powders of aluminium oxide and carbonyl iron, nanocrystallised preliminarily, with the size of crystallites around 10 nm [24].

2. In order to preserve nanosized structure of pressing and to avoid grain growth during sintering there was used the method of dry isostatic pressing (P = 488 MPa) under the action of ultrasonic waves at a power of W = 0, 1 and 3 kW [25]. Ultrasonic effect destroys agglomerates and activates particles during pressing. This provides efficient consolidation of material at the initial stages of pressing, and also 3-4 times reduction of elastic aftereffect in the pressing. As a result, internal stresses inside the pressing are reduced, its strength is increased,

more regular density of a compact is provided, grain growth during the subsequent sintering is retarded.

After the milling of the mechanical mixture of the initial powder for 40 min, the size of particles in the mixture according to the transmission electron microscopic data is equal to 30–200 nm and does not depend on the iron content in the initial powder. According to the X-ray powder diffraction data, during the mechanochemical activation of the initial powder, boehmite undergoes phase transformation into the γ-Al_2O_3 modification. The activation of the metastable modifications of aluminum oxide under the chosen conditions results in a broadening of the X-ray lines [26]. In this case, the average size of coherent scattering regions of iron remains unchanged and is approximately equal to 10 nm. The analysis of the X-ray photoelectron spectra demonstrates that the surface of particles of the Al_2O_3 + 1 wt % Fe and Al_2O_3 + 20 wt % Fe mixtures does not contain iron atoms; i.e., grains of the metallic phase are located below the Al_2O_3 oxide layer. This favors the retention of the phase state of grains and protects the metallic phase from the oxidation in air upon heating to a temperature of ~1346 K [27].

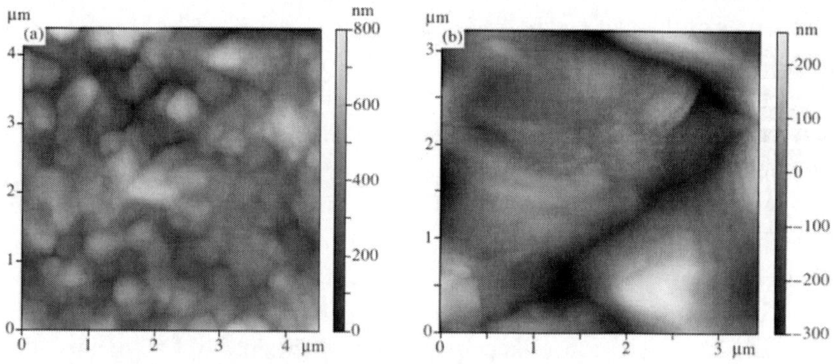

Figure 11. AFM (atomic force microscopy) images of the cleavage surface of a sample with 15 wt % Fe compacted at W = 0 kW and sintered at (a) T = 1673 KC and time t = 0.5 h and (b) T = 1723 K and t = 2 h.

The thermal sintering of the compacted samples was carried out in a vacuum chamber with a residual pressure of $1.33 \cdot 10^{-2}$ Pa to a temperature of 1673 K at a heating rate of 10 K/min and calcined for 30 min, followed by cooling with the furnace. The process is accompanied by the phase transition of the metastable aluminum oxides to the high-temperature α-phase and the formation of the $FeAl_2O_4$ spinel; in this case, the width of the reflections of the α-Fe phase decreases as a result of the recrystallization process. The size of coherent scattering regions of the aluminum oxide (calculated from the broadening of the X-ray line) is equal to 60–160 nm (Table 4) and is almost independent of the iron content in the initial powder. Samples, pressed at ultra-sound exposure capacity 1 kW, feature the smallest size of crystallites. The content of the spinel phases did not exceed 3 wt %.

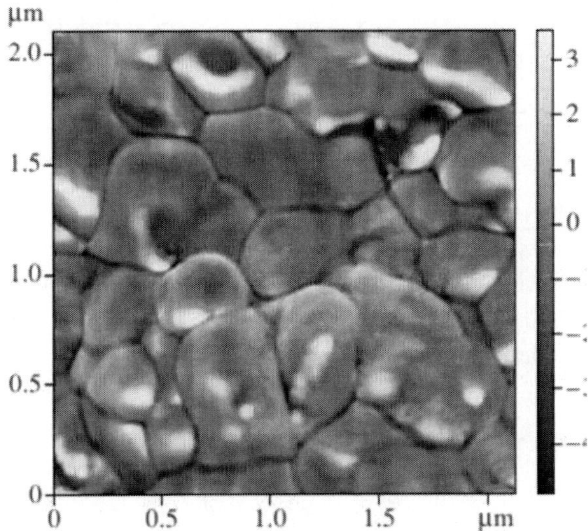

Figure 12. AFM phase-contrast image of a sample with 10 wt % Fe. Bright regions correspond to a softer metallic phase. AFM topography images of the cleavage surface of a sample with 15 wt % Fe compacted at $W = 0$ kW and sintered at (a) $T = 1400°C$ and time $t = 0.5$ h.

According to atomic-force microscopy investigations in the topography mode, the cermets contain grains smaller than 300 nm in size. The sizes of Al_2O_3 grains are controlled by the holding temperature and time and are independent of the iron content. Figure 11a shows the surface morphology of the cleavages of the samples annealed at 1673 K, which is typical of all concentrations, and Figure 11b shows the surface morphology of the sample annealed at 1723 K for 2 h.The use of the phase-contrast mode (sensitive to a change in the local elastic properties of the surface) allows us to reveal metal grains in the form of granules (30–80 nm in size) with a shape similar to spherical against the background of a polycrystalline ensemble of Al_2O_3 grains (Figure 12). Examination of a large area of the cleavage of the cermet demonstrates that the metal grains are uniformly distributed without clustering or formation of whisker-type structures.

Table 4.

N	Phase composition, wt %	W, kW	Sintering temp./ sintering time, T/t, K/h	R_{Al2O3}, nm	D,cm²/s, T = 3.86 K	h
2	Al_2O_3:10%Fe	0	1400/0.5	94	$3.26 \cdot 10^{-2}$	2
3	Al_2O_3:15%Fe	0	1400/0.5	79	$2.30 \cdot 10^{-2}$	1
4	Al_2O_3:15%Fe	0	1450/2	>300	$2.27 \cdot 10^{-1}$	1
5	Al_2O_3:5%Fe	1	1400/0.5	74	$3.2 \cdot 10^{-2}$	3
6	Al_2O_3:10%Fe	1	1400/0.5	68	$2.86 \cdot 10^{-2}$	2
7	Al_2O_3:15%Fe	1	1400/0.5	65	$2.65 \cdot 10^{-2}$	1
8	Al_2O_3:5%Fe	3	1400/0.5	93	$5.60 \cdot 10^{-2}$	3
9	Al_2O_3:10%Fe	3	1400/0.5	87	$3.38 \cdot 10^{-2}$	2

The Table 4 gives the structural parameters of the samples, the technological conditions, the diffusion coefficients measured at T = 3.86 K, and the exponents in the temperature dependences $D(T) \sim T^h$ for the temperature range 2.3–3.86 K.

When analyzing the experimental results, we revealed the following three main features:

a) As the fraction of the metallic phase increases, the diffusion coefficient at T = 3.86 K decreases.

b) The diffusion coefficients are very low (they are two orders of magnitude lower than those calculated using the $D(R) \sim R$ dependence in the geometrical-scattering range; (see Figure 3).

c) The temperature dependence has the form $D(T) \sim T^h$ with an exponent $h = 1 \div 3$, which decreases as the amount of the metallic phase increases.

Obviously, as the fraction of the metallic phase (whose thermal conductivity is higher than that of corundum at liquid-helium temperatures) increases, the thermal conductivity of the composite and, hence, the diffusion coefficient of phonons in a macroscopic dielectric sample should increase [28]. However, in our studies, the diffusion coefficient of NPs decreases with increasing fraction of the metallic phase.

On the other hand, in Section II, we presented a theoretical model that describes the transport of NPs in nanoceramics and is based on the concept of randomly distributed spherical shells, which simulate grain boundaries and have elastic properties other than those of grains. We showed that the restructuring of the phonon spectrum of ceramics caused by phonon scattering by nanoobjects at $qR \sim 1$ can lead to various temperature dependences of the diffusion coefficient of

NPs and to a significant decrease in its absolute values. Unfortunately, this model cannot be applied to the cermets studied in this work. Let us consider a simple example.

Figure 3 shows the dependence of the diffusion coefficient on the parameter $x= qR_{Al2O3} \sim T$ at various values of the parameter $x_r= \omega_r R_{Al2O3}/v$, where ω_r is the resonance frequency of NP scattering by shells of radius R. Parameter x_r is the main model parameter: a change in its value results in a wide spectrum of the diffusion coefficient and its temperature dependence at qR of several units (see Figure 3). Nevertheless, the diffusion coefficient measured in this work at T = 3.86 K ($D \leq 0.05$ cm^2/s) can only exist near the gap edge (where the diffusion coefficient increases exponentially with temperature), which is in conflict with the detected temperature dependences $D(T) \sim T^h$, where h = 1÷3. The presence of fine inclusions (metallic fraction) in the system changes the gap (minimum) position but cannot qualitatively change the behavior of the $D(T)$ dependence; that is, for the results obtained to be explained, we have to assume an additional mechanism. For example, metallic grains can be considered as the centers of NP entrapment. The fact that metallic nanofragments can serve as NP traps follows from the relation between the time of electron–phonon interaction $\tau_{e\text{-}ph}$ and the time τ_R of the ballistic NP path in the bulk of an iron grain, $\tau_R = R_{Fe}/vc$ (where c is the coefficient of exit of NPs from a grain, $10^{-2} < c < 1$). If

$$\tau_{e-ph} << \tau_R \,, \tag{12}$$

metallic grains can serve as NP traps. $ql_{el} >> 1$ [29], we have

$$\tau_{e-ph} = \frac{6\rho v}{\pi n m V_F} \cdot \frac{1}{\omega}$$

where n and ρ are the iron concentration and density, respectively; m is the electron mass, and V_F is the Fermi velocity. Estimates demonstrate that, under the experimental conditions, inequality (12) is fulfilled at $\omega \geq 10^{12}$ (c = 0.1). However, $\omega = \frac{kT}{\hbar} = 5 \cdot 10^{11} s^{-1}$ at T = 4 K.

Thus, the NPs with a frequency $\omega > (2\text{-}3)\frac{kT}{\hbar}$ from the phonon spectral region corresponding to the maximum spectral density of the Planck distribution at $T \geq 4$ K can interact with electrons in the metal.

The presence of factors that hinder the exit of Nps from metallic grains can be indicated by the fact that the acoustic impedances of the metal and corundum are significantly different. If metallic-phase grains are surrounded by an $FeAl_2O_4$ spinel layer, the ratio of the acoustic impedances of iron and the spinel is about 2/1, which should hinder the exit of NPs from a grain to the polycrystalline Al_2O_3 matrix.

A detailed analysis of the effect of the structure of grain boundaries on NP scattering is beyond the scope of this work. We can correctly determine the contribution of NP scattering by grain boundaries under our experimental conditions only in the geometrical-scattering range ($qR >> 1$), where $D \sim Rvf_\omega$ in dense dielectric ceramics; here, f_ω is the scattering probability of NPs that have frequency ω and pass through grain boundaries [16]. In this case, a grain boundary is simulated as a flat layer between material fragments (grains). In the case of $qR \sim 1$, the consideration of this problem in a multiphase system requires the introduction of a large number of interdependent parameters [6]. In [9], we showed that, as the temperature or time of sintering of ceramics decreases (in order to form a material with a small grain size), the grain-boundary thickness increases and the grain-boundary acoustic impedance and elasticity decrease. As a result, the absolute values of the diffusion coefficient decrease and the temperature dependence *D(T)* changes. A decrease in the elasticity of the grain-boundary material as compared to that of the matrix-grain material increases the efficiency of the resonance scattering of NPs in the range $qR \sim 1$ [6], which is indicated by the weakening of the *D(R)* dependence even at $qR_{Al2O3} < 20$ (see Figure 3).

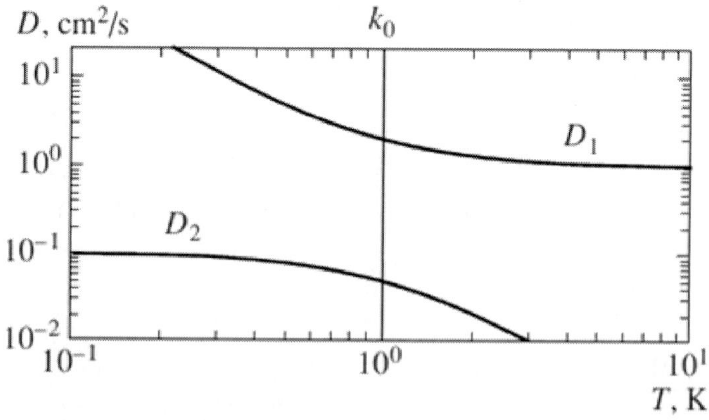

Figure 13. Dispersion of diffusion coefficients *D(T)* calculated by Eqs. (3) and (4).

As follows from the experimental results, the diffusion coefficient of NPs under conditions of an effective electron–phonon interaction (NP entrapment by metallic inclusions) is one to two orders of magnitude lower than that in an analogous polycrystalline Al_2O_3 matrix, which mainly specifies the transport properties of Nps under our experimental conditions.

The propagation of NPs in the presence of trapping centers was considered in [14,15], where experimental results on the propagation of NPs in solid solutions with $Y_{3-x}(Er,Ho)_xAl_5O_{12}$ paramagnetic centers were analyzed. According to those results, the propagation of a δ-like thermal pulse in an infinite one-dimensional medium is described by two diffusion coefficients, which characterize “fast” and “slow” processes.

For the spatial Fourier transform of the temperature distribution at time t, the author of [14,15] obtained the expression

$$S(t,k) = S(0)(A_1(k)e^{-k^2 D_1(k)t} + A_2(k)e^{-k^2 D_2(k)t})$$

in which the diffusion coefficients are determined from the formula

$$\frac{D_{1,2}(k')}{D_0}k'^2 = (k_0^2 + k^2 \pm \sqrt{(k_0^2 + k'^2)^2 - 4k'^2 k_0^2 C})/2$$

Here, $k'=kL$, where L is the coordinate of the point of temperature measurement (sample length), and D_0 is the diffusion coefficient determined by only elastic phonon scattering in the matrix material. The dispersion of the diffusion coefficients is shown in Fig.13. Quantity k_0 is seen to be the most important parameter of the problem. At $k' < k_0$ (for definiteness, we assume $D_1 > D_2$), the temperature-field distribution is only contributed by slow processes, which are characterized by effective diffusion coefficient $D_2 = CD_0$. Quantities C and k_0 are physical parameters of the system: C determines the fraction of the contribution of the phonon subsystem (c_{ph}) to the total specific heat of the sample,

$$C = c_{ph}/(c_{ph} + c_{tr}) < 1,$$

where c_{tr} is the specific heat of trapping centers, and the quantity

$$k_0 = \sqrt{2t_0/\tau_{e\text{-}ph}} = \sqrt{L^2/D_0 \cdot \tau_{e\text{-}ph}}$$

is proportional to the number of phonons that inelastically interact with trapping centers (are trapped) within the time of passage of length L along the sample. Since the condition $k_0>>1$ is met, we can neglect fast processes with a characteristic diffusion coefficient $D_1 = D_0$ and can write the effective diffusion coefficient of the process in the form

$$D_{eff} = \frac{D_0}{1 + c_{tr} / c_{ph}}$$

In the cermets under study, trapping centers are represented by metallic inclusions, since condition (12) is met when a phonon wave packet passes through the sample. We take into account the specific heats of the matrix and metallic inclusions and write the dependence of D_{eff} on the iron weight concentration p in the form

$$D_{eff} = \frac{D_0}{1 + \dfrac{pc_{Fe}}{(1-p)c_{Al_2O_3}}} \qquad (13)$$

Since the electronic specific heat of iron at a concentration of about 10 wt % and liquid-helium temperatures is approximately two orders of magnitude higher than that of the phonon subsystem (c), this expression Al_2O_3 demonstrates that D_{eff} decreases as $D \sim 1/p$ as the iron concentration increases. At liquid-helium temperatures, we have $c_{el}/c_{ph} \sim T^{-2}$; therefore, we obtain $D_{eff} \sim D_0 T^2$. In dielectric ceramics, D_0 is controlled by elastic phonon scattering by grain boundaries; that is, it behaves according to the model in [6]. As is seen from Figure 3, the detected temperature dependence $D \sim T^h$ (h = 1÷3) can be obtained for phonons with qR of several units (to the right of the gap in the phonon spectrum), which corresponds to the experimental data.

Figure 14 shows the experimental and calculated $D(T)$ dependences, and their numerals correspond to those in the table. Numerals with primes indicate the temperature dependences of the diffusion coefficient in the structure of a polycrystalline Al_2O_3 matrix that were calculated according to [6] without regard for the entrapment of NPs by fine iron inclusions. The calculated dependences that describe the experimental data were obtained using Eq. (13), which takes into account the ratio of the specific heats in the given temperature range. As the iron-fraction concentration increases, the absolute values of D at T = 3.86 K decrease and the temperature dependence $D(T)$ weakens. According to [6], this behavior indicates an increase in resonance frequency ωr , which is caused by NP scattering by the structural fragments in the sample under study. Curves 4' and 4 in Fig. 14 correspond to a sample with larger grains in the polycrystalline matrix (*T/t* = 1450/2, where t is the sintering time) and to the fraction of nanodispersed iron that is identical to that for curves 3' and 3 (which is 15 wt %). In this case, the diffusion coefficient increases by an order of magnitude because of the weakening of the resonance character of NP scattering by the larger grains of the Al_2O_3 fraction.

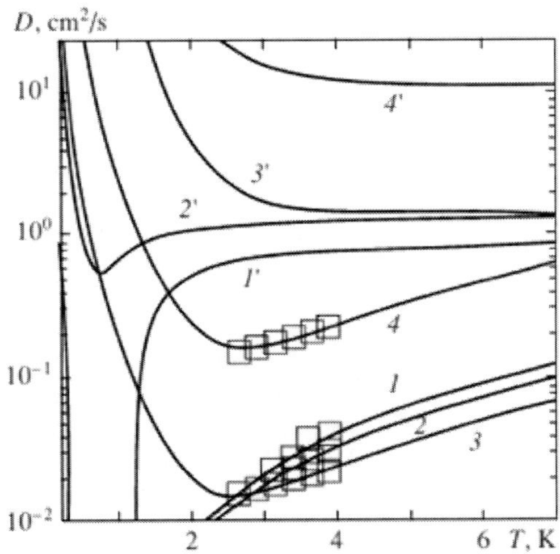

Figure 14. Temperature dependences *D(T)* for the cermet samples synthesized without an ultrasonic action at the stage of compacting: (1'–4') $D_0(T)$ dependences calculated for a polycrystalline matrix without regard for the fraction of iron (1–4) dependences calculated by Eq. (13), and (squares) experimental data (numerals at the curves correspond the numerals in the Table 4).

An ultrasonic action is known to change the dispersity of a ceramic matrix [30, 31]. As follows from the results shown in Figures 15 and 16, these structural changes also affect the diffusion coefficient. For example, an ultrasonic action at a power of W = 3 kW initiates recrystallization during sintering of samples, which leads to a decrease in the imperfection of grain boundaries and to a simultaneous increase in the grain size. This affects the behavior of the diffusion coefficient (Figure 15), which is described by higher values of x_r and R. In contrast, at W = 1 kW, the Al_2O_3 grain size decreases as compared to the case of W = 0 at the stage of compacting, which results in a decrease in x_r. Figure 16 shows the results of such changes in the structure of a composite with 5 wt % Fe.

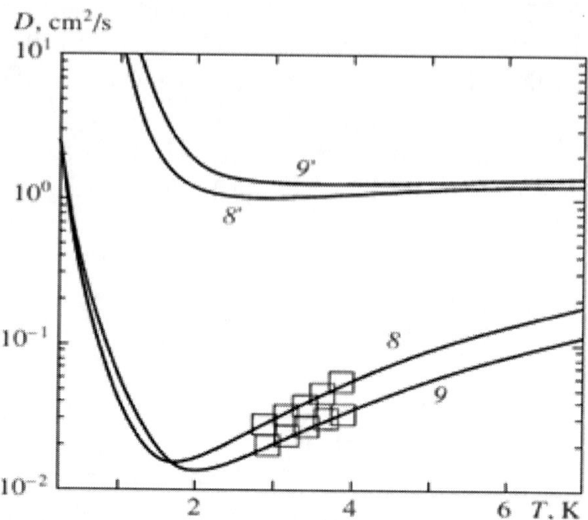

Figure 15. Temperature dependences *D(T)* for the cermet samples synthesized during the action of 3 kW ultrasound: (8',9') $D_0(T)$ dependences calculated for a polycrystalline matrix without regard for the fraction of iron [6], (8, 9) dependences calculated by Eq. (13), and (squares) experimental data (numerals at the curves correspond to the numerals in the Table 4).

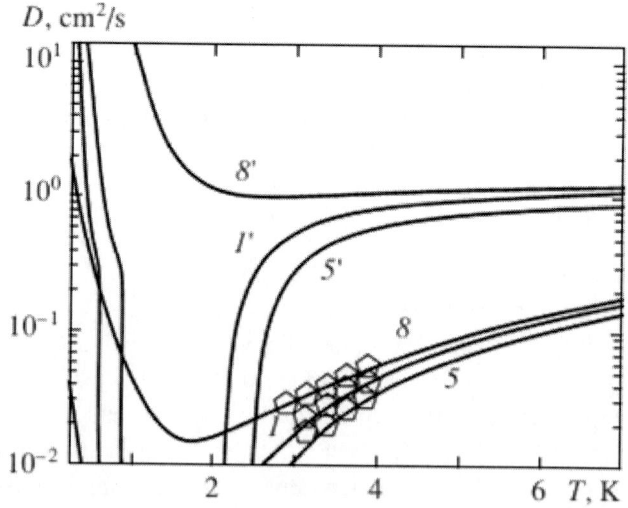

Figure 16. Temperature dependences *D(T)* for the cermet samples with 5 wt % Fe synthesized during the action of ultrasound at a power of (1, 1') 0, (5, 5') 1, and (8, 8') 3 kW. (Symbols) experimental data (numerals at the curves correspond to the numerals in the Table 4).

Conclusion

The research represents the results of experimental studies of transporting behaviour of phonons in mono-phase and multi-phase dielectric ceramics obtained by the method of compaction. It was found that if the average particle size (*R*) of ceramics decreases then the behaviour of phonon diffusion ratio (at temperatures of liquid helium) changes qualitatively before reaching some critical value (~100nm). Linear relation D (R) transforms to more strong (even exponential), at this the change of derivative sign *dD/dT* is observed - if at higher R the temperature relation of diffusion ratio behaves normally ($dD/dT<0$), then in nano-ceramics it increases abnormally with the temperature growth.

With analysis of experimental results, theoretical model which explains these anomalies was worked out. According to this model, there may appear resonance phonon scattering at grain boundaries in nanoceramics. This can lead to strong phonon spectrum renormalization and, at some conditions, creation of range of forbidden values at low-frequency part of phonon spectrum of a system (gap). Location and width of the gap is determined by the size and elastic properties of grains and inter-phase boundaries, i.e. conditions of ceramics synthesis. Presence of systems with the gap in the phonon spectrum of the system is a requirement for creation of thermo-crystals - nanostructured systems for control and direction of heat flows.

Also additional methods of direction of heat flows in ceramics were discussed. There was studied the influence of nanosized metal inclusions in non-conductive ceramic matrix on the coefficient of phonon diffusion. It was shown that at temperatures of the experiment ($T \leq 4K$), when electronic specific heat of these inclusions is much more than that of phonons, nanosized metal inclusions act as efficient centers of phonon capture. This can result in great (up to several degrees) decrease of diffusion ratio.

It was shown that due to significant difference of thermal properties of these materials, there appeared technological problems during synthesis of ceramic composites (oxide + metal) connected with preservation of sizes of metal inclusions and prevention of metal oxidation when sintering. There was suggested a new method of the powder preparation which allowed to solve those problems.

Thus the nanoceramics synthesis by means of compaction is considered as one of the ways to create systems with the gap in phonon spectrum. This method allows to obtain systems with forbidden gap of a determined width in given frequency range, which is essential for thermal crystals synthesis. Simplicity of obtaining of such systems comparing to phonon lattice creation allows to decrease significantly the price/quality ratio during production.

This work was supported by the Russian Foundation for Basic Research (project no. 15-07-00704 and 16-07-00592).

References

[1] Maldovan, *Phys. Rev. Lett.* 110, 025902 (2013).

[2] M. Maldovan, *Nature* 110, 209 (2013).

[3] E. I. Salamatov, in Proceedings of the 11th International Conference on Phonon Scattering in Condensed Matter (PHONONS 2004), edited by A. Kaplyanskii, A. Akimov, and V. Bursian (Wiley-VCH Verlag GmbH, Weinheim, Germany, 2004), p. 2971.

[4] Y. N. Barabanenkov, V. V. Ivanov, S. N. Ivanov, E. Salamatov, A. V. Taranov, E. N. Khazanov, and O. L. Khasanov, *JETP* 102, 114 (2006).

[5] R. von Gutfeld and A. Nethercot, Jr., *Phys. Rev. Lett.* 12, 641 (1964).

[6] V. V. Ivanov, E. Salamatov, A. V. Taranov, and E. N. Khazanov, *JETP* 106, 288 (2008).

[7] V. V. Ivanov, S. N. Ivanov, O. V. Karban, et al., *Inorganic Materials* 40, 1233 (2004)

[8] V. V. Ivanov, E. Salamatov, A. V. Taranov, and E. N. Khazanov, *JETP* 110, 34 (2010).

[9] V. V. Ivanov, V. R. Khrustov, S. N. Paranin, A. I. Medvedev, A. Shtol'ts, O. Ivanova, and A. Nozdrin, Glass Phys. Chem. 31, 465 (2005).

[10] O. L. Khasanov, E.S. Dvilis, *Advances in Applied Ceramics* 107, 135 (2008).

[11] E. Salamatov, *Phys. Status Solidi* B 244, 1895 (2007).

[12] E. Salamatov, *Phys. Status Solidi* B 246, 92 (2009).

[13] E. Salamatov, A.Taranov, and E. Khazanov, *J. Appl. Phys.* 114, 154305 (2013).

[14] E. I. Salamatov, *Phys. Solid State* 44, 978 (2002).

[15] E. I. Salamatov, *Phys. Solid State* 45, 725 (2003).

[16] Yu. N. Barabanenkov, V. V. Ivanov, S. N. Ivanov, et al., *JETP* 92, 474 (2001).

[17] A. P. Zhernov, E. I. Salamatov, and E. P. Chulkin, *Phys.Status Solidi B* 168, 81 (1991).

[18] A. P. Zhernov, E. Salamatov, and E. P. Chulkin, *Phys. Status Solidi B* 165,355 (1991).

[19] V. V. Ivanov, S. N. Paranin, and V. R. Khrustov, *Phys. Met. Metallogr.* 94 (Suppl. 1), S98 (2002).

[20] J. L. Guichard, O. Tillement, and A. Mocellin, *J. Eur. Ceram. Soc.* 18, 1743 (1998).

[21] *D. Osso,* O. Tillement, G. Le Caer, A. Mocellin, *Journal of Materials Science* 33, 3109 (1998).

[22] E. Gaffet, F. Bernard, J.-C. Niepce, F. Charlot, et al., *J. Mater. Chem.* 9, 305 (1999)

[23] D. R. Clarke, S. R. Phillpot, *Materials today* 6, 22 (2005).

[24] E. P. Elsukov, G. A. Dorofeev, A. I. Ul'yanov, A. V. Zagainov, A. N. Maratkanova, *Phys. Met. Metallogr.* 91, 258 (2001)

[25] O. Karban, E. Salamatov, A. V. Taranov, E. N. Khazanov, and O. L. Khasanov, *JETP* 108, 661 (2009).

[26] J. R. Groza. *Int.J. Powder Metall.* 35, 59 (1999).

[27] Coquay, P., Laurent, Ch., Peigney, A., Quénard, et al., *Hyperfine Interact.* 130, 275 (2000).

[28] D. V. Liu and W. Y. Tuan, *Acta Mater.* 44, 813 (1996).

[29] A. B. Pippard, *Philos. Mag.* 46, 1104 (1955).

[30] O. L. Khasanov, O. V. Karban, E.S. Dvilis, Key *Engineering Materials* 264-268, 2327 (2004).

[31] O. Karban, O. Kanunnikova, E. Khazanov, et al., *Ceramics International* 39, 497 (2013).

In: Ceramic Materials
Editor: Jacqueline Perez

ISBN: 978-1-63485-965-3

Chapter 3

ANALYTICAL MODEL OF THERMAL STRESSES IN MULTI-COMPONENT MATERIALS WITH CYLINDRICAL PARTICLES

Ladislav Ceniga*
Institute of Materials Research, Slovak Academy of Sciences,
Košice, Slovak Republic

Abstract

The paper deals with analytical modelling of thermal stresses in a multi-particle-matrix system with isotropic cylindrical particles which are periodically distributed in an isotropic infinite matrix. This multi-particle-matrix system represents a model system which is applicable to real two-component materials of with precipitates of a cylindrical shape. The thermal stresses as functions of microstructural parameters (particle volume fraction, particle radius, inter-particle distance) originate during a cooling process as a consequence of the difference in thermal expansion coefficients of the cylindrical particle and the matrix. The analytical modelling is based on an application of suitable mathematical techniques on fundamental equations of mechanics of solid elastic continuum. The mathematical techniques thus result in analytical solutions for both the cylindrical particle and the infinite matrix. Finally, numerical values of the thermal stresses in a real two-component material with cylindrical precipitates are determined.

*E-mail addresses: lceniga@yahoo.com, ladislavceniga65@gmail.com, lceniga@saske.sk, lceniga@imr.saske.sk

PACS: 46.25.Cc, 46.25.Hf, 46.70.Lk

Keywords: analytical modelling, thermal stress, composite material

1. Introduction

Thermal stresses which originate as a consequence of different thermal expansion coefficients of components of multi-component materials represent an important phenomenon in multi-component materials. These stresses are usually investigated by computational and experimental methods [1]–[7] are still of interest to materials scientists and engineers.

This chapter represents continuation of the author's book [8] which deals with the analytical modelling of the thermal stresses in two-component materials of a precipitate-matrix type with isotropic *spherical* precipitates (isotropic particles) and an isotropic matrix. In contrast to [8], the analytical model of the thermal stresses in this chapter is determined for an isotropic matrix with *cylindrical* particles (=whiskers).

2. Cell Model

With regard to the analytical modelling [8], real two-component materials of finite dimensions are replaced by an infinite multi-particle-matrix system (see Fig. 1). The multi-particle-matrix system represents a model system. The model system is represented by isotropic cylindrical particles which are periodically distributed in an isotropic infinite matrix. The infinite matrix is imaginarily divided into identical cubic cells with a central cylindrical particle. The cubic cell represents such part of the multi-particle-matrix system which is related to one cylindrical particle. As presented in the cubic cells *1234*, *5678*, *89 10 11*, the central cylindrical particles exhibit mutually different orientation.

This 'cell approach' is used within mathematical procedures applied to analytical modelling of periodic model systems with an infinite matrix [9]–[20]. The thermal stresses are thus investigated in the cubic cell. On the one hand, the influence of the thermal stresses of all cells on the thermal stresses in a certain cell as well as the influence of the thermal stresses of neighbouring cells on the thermal stresses in a certain cell can be hardly expressed by an exact mathematical solution. On the other hand, the analytical model of the thermal stresses which is determined within a cell is generally considered to represent more than

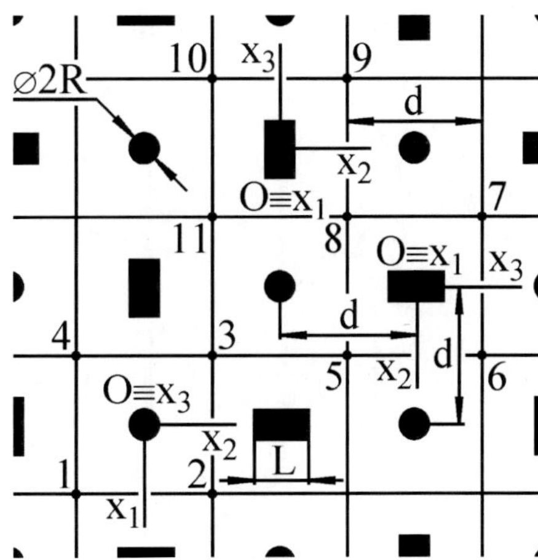

Figure 1. The multi-particle-matrix system imaginarily divided into identical cubic cells with a central cylindrical particle in the point O of the Cartesian system $(Ox_1x_2x_3)$, where R and L are radius and length of the cylindrical particle, respectively, and a dimension of the cubic cell is identical to the inter-particle distance d. As presented in the cubic cells *1234*, *5678*, *89 10 11*, the central cylindrical particles exhibit mutually different orientation.

sufficient mathematical solution for materials scientists and engineers. Using final mathematical formulae for the thermal stresses, materials scientists and engineers can then determine numerical values of the thermal stresses for a real two-component material.

Finally, as presented in [9, 20], to obtain analytical solutions, real two-component materials with finite dimensions and with aperiodically distributed particles (precipitates) are then replaced by a model system which consists of an infinite matrix with periodically distributed particles of a defined shape (e.g. spherical particles). Additionally, the case when an infinite matrix is considered within the analytical modelling is of particular interest for the mathematical simplicity of analytical solutions. As analysed in [20], such analytical solutions are assumed to exhibit sufficient accuracy due to the size of material components (e.g. precipitates) which is relatively small in comparison with the size of macroscopic material samples, macroscopic structural elements, etc.

The inter-particle distance d, the particle radius R, the particle length L, the particle volume fraction v, which represent parameters of the cubic cells,

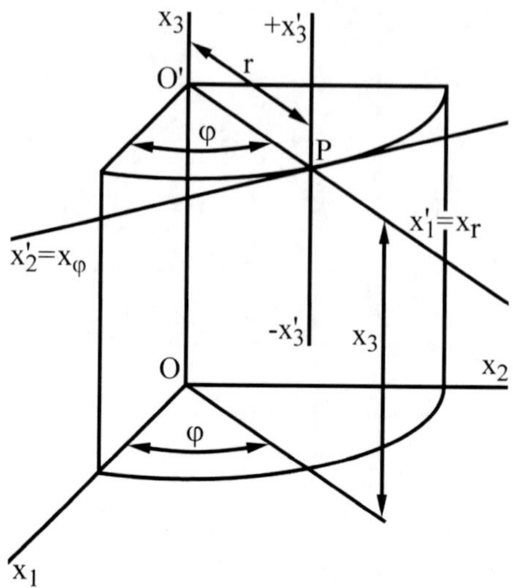

Figure 2. The axes $x_1' = x_r = O'P$ and $x_2' = x_\varphi \parallel x_1x_2$, $x_3' = x_3$ defining radial and tangential directions regarding the Cartesian system $(Px_1'x_2'x_3')$, respectively; the arbitrary point P with a position determined by the cylindrical coordinates $[r, \varphi, x_3]$ regarding the Cartesian system $(Ox_1x_2x_3)$ (see Fig. 1); the axes $+x_3'$ and $-x_3'$ regarding the point P. The point O is a centre of the cylindrical particle, and x_2', x_3' are tangents to the surface of a cylinder with the radius $r = \left|\overline{O'P}\right|$ representing length of the abscissa $|O'P|$.

are microstructural parameters of two-component materials of the precipitate-matrix type with cylindrical particles, where v has the form

$$v_p = \frac{V_p}{V_c} = \frac{\pi R^2 L}{d^3} \in (0, v_{pmax}\rangle\,, \quad v_{pmax} = \frac{\pi}{4}, \tag{1}$$

where $V_p = \pi R^2 L$ and $V_c = d^3$ is volume of the spherical particle, the spherical pore and the cubic matrix, respectively. The maximum value v_{pmax} results from the condition $R = d/2$ and $L = d$.

In case of a real two-component material of the precipitate-matrix type, these microstructural parameters are obtained by experimental techniques. As mentioned above, the thermal stresses are investigated within the cubic cell with the cylindrical particle, and thus represent functions of these microstructural parameters. Additionally, the surface of the cubic cell defines positions for which one of boundary conditions for the infinite matrix is determined (see the distance r_c, Eqs. (2), (65)).

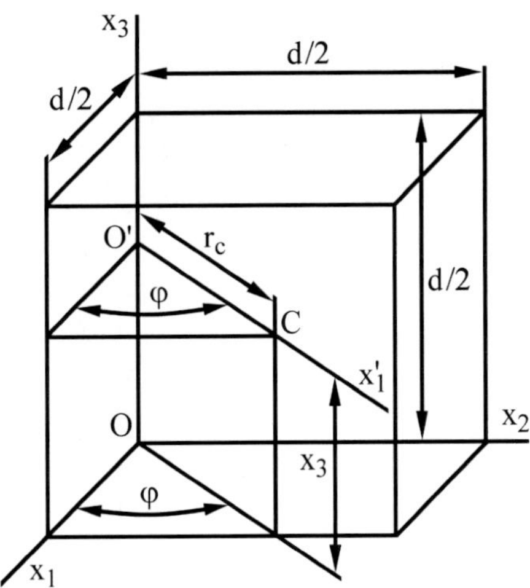

Figure 3. The point C as a point of intersection of the axis x_1', representing a radial direction, with a surface of one eighth of the cubic cell (see Fig. 1), i.e. for $\varphi \in \langle 0, \pi/2 \rangle$, where $r_c = \left|\overline{O'C}\right|$ is given by Eq. (2).

The thermal stresses, as functions of the cylindrical coordinates $[r, \varphi, x_3]$, are derived in the Cartesian system $(Px_1'x_2'x_3')$ (see Fig. 2). The cylindrical coordinates (r, φ, x_3) are used due to the cylindrical shape of the particles of the multi-particle-matrix system. The axes x_1' and x_2', x_3' represent radial and tangential directions, respectively, and $r = \left|\overline{O'P}\right|$. Due to the symmetry of the cubic cell as well as due to the isotropy of the multi-particle-matrix system, the thermal stresses are sufficient to be investigated within one eighth of the cubic cell, i.e. for $\varphi \in \langle 0, \pi/2 \rangle$ and $x_3 \in \langle 0, d/2 \rangle$, where $r \in \langle 0, R \rangle$ for the spherical particle and $r \in \langle R, r_c \rangle$ for the cell matrix, respectively.

The distance $r_c = r_c(\varphi) = \left|\overline{O'P}\right|$ (see Fig. 3) has the form

$$r_c = \frac{d}{2\cos\varphi}, \quad \varphi \in \left\langle 0, \frac{\pi}{4} \right\rangle; \qquad r_c = \frac{d}{2\sin\varphi}, \quad \varphi \in \left(\frac{\pi}{4}, \frac{\pi}{2} \right\rangle. \tag{2}$$

3. Fundamental Equations

The analytical modelling presented in this chapter results from fundamental equations of solid continuum mechanics. These fundamental equations are represented by the Cauchy's equations, the equilibrium equations and the Hooke's law for an isotropic elastic solid continuum. The Cauchy's equations and the equilibrium equations for the infinitesimal cylindrical cap in the arbitrary point P with a position determined by the cylindrical coordinates $[r, \varphi, x_3]$ regarding the Cartesian system $(Ox_1x_2x_3)$ are determined for the Cartesian system $(Px_1'x_2'x_3')$ (see Fig. 2). The Cauchy's equations which represent geometric equations define relationships between strains and displacements of an infinitesimal part of a solid continuum. The equilibrium equations which are related to the axes x_1', x_2', x_3' (see Fig. 2) are based on a condition of the equilibrium of forces which act on sides of this infinitesimal part.

The infinitesimal spherical cap in the arbitrary point P is described by the dimensions dr, $rd\varphi$, dx_3 along the axes x_1', x_2', x_3' of the Cartesian system $(Px_1'x_2'x_3')$, respectively. The axis x_1' represents a normal of the surfaces S_r and S_{r+dr} with the surface area $A_r = r\, d\varphi\, dx_3$ and $A_{r+dr} = (r+dr)\, d\varphi\, dx_3$ at the radii r and $r+dr$, respectively. Due to the symmetry of the multi-particle-matrix system, this infinitesimal cylindrical cap exhibits the axial displacement u_3' along the axis x_3' and the radial displacement u_1' along the axis x_1' (see Fig. 2) as similar to infinitesimal spherical cap analysed in [21]. Additionally, due to the symmetry of the multi-particle-matrix system, the displacement u_3' along the axis x_3' is not a function of the variable $\varphi \in \langle 0, \pi/2)$, and then $(\partial u_3'/\partial\varphi) = 0$. s Accordingly, the Cauchy's equations for the radial strain ε_{11}' along the axis x_1', tangential strain ε_{ii}' and the shear strains $\varepsilon_{1i}' = \varepsilon_{i1}'$ along the axis x_i' $(i = 2,3)$ are derived as [22]–[24]

$$\varepsilon_{11}' = \frac{\partial u_1'}{\partial r}, \tag{3}$$

$$\varepsilon_{22}' = \varepsilon_{33}' = \frac{u_1'}{r}, \tag{4}$$

$$\varepsilon_{33}' = \frac{\partial u_3'}{\partial x_3}, \tag{5}$$

$$\varepsilon_{12}' = \varepsilon_{21}' = \frac{1}{r}\frac{\partial u_1'}{\partial \varphi}, \tag{6}$$

$$\varepsilon_{13}' = \varepsilon_{31}' = \frac{1}{2}\left(\frac{\partial u_1'}{\partial x_3} + \frac{\partial u_3'}{\partial r}\right), \tag{7}$$

Consequently, the equilibrium equations for the radial stress σ'_{11} acting along the axis x'_1, tangential stresses σ'_{ii} and the shear stresses $\sigma'_{1i} = \sigma'_{i1}$ acting along the axis x'_i ($i = 2,3$) have the forms [22]–[24]

$$\sigma'_{11} - \sigma'_{22} + r\left(\frac{\partial\sigma'_{11}}{\partial r} + \frac{\partial\sigma'_{13}}{\partial x_3}\right) + \frac{\partial\sigma'_{12}}{\partial\varphi} = 0, \tag{8}$$

$$2\sigma'_{12} + r\,\frac{\partial\sigma'_{12}}{\partial r} + \frac{\partial\sigma'_{22}}{\partial\varphi} = 0, \tag{9}$$

$$\sigma'_{13} + r\left(\frac{\partial\sigma'_{13}}{\partial r} + \frac{\partial\sigma'_{33}}{\partial x_3}\right) = 0. \tag{10}$$

With regard to the shear strain $\varepsilon'_{23} = \varepsilon'_{32} \propto [(\partial u'_2/\partial x_3) + (1/r) \times (\partial u'_3/\partial\varphi)]$ in the Cartesian system $(Px'_1x'_2x'_3)$ (see Fig. 2) [22]–[24], we get $\varepsilon'_{23} = 0$ due to $u'_2 = 0$ and $(\partial u'_3/\partial\varphi) = 0$, where u'_2 and u'_3 represent displacements of the infinitesimal cylindrical cap along the axes x'_2 and x'_3 (see Fig. 2), respectively, i.e. along tangential directions. Similarly, with regard to the Hooke's law for an isotropic elastic solid continuum, i.e. $\varepsilon'_{23} = s_{44}\sigma'_{23}$ (see Eq. (16)) [22], we get $\sigma'_{23} = 0$.

Finally, with regard to $\varepsilon'_{23} = 0$, $\sigma'_{23} = 0$, the Hooke's law for an isotropic elastic solid continuum is derived as [22]–[24]

$$\varepsilon'_{11} = s_{11}\sigma'_{11} + s_{12}\left(\sigma'_{22} + \sigma'_{33}\right), \tag{11}$$

$$\varepsilon'_{22} = s_{12}\left(\sigma'_{11} + \sigma'_{33}\right) + s_{11}\sigma'_{22}, \tag{12}$$

$$\varepsilon'_{33} = s_{12}\left(\sigma'_{11} + \sigma'_{22}\right) + s_{11}\sigma'_{33}, \tag{13}$$

$$\varepsilon'_{13} = s_{44}\sigma'_{13}, \tag{14}$$

$$\varepsilon'_{12} = s_{44}\sigma'_{12}. \tag{15}$$

The elastic moduli s_{11}, s_{12}, s_{44} which represent functions of the Young's modulus E and Poisson's ratio μ have the forms [22]–[24]

$$s_{11} = \frac{1}{E}, \quad s_{12} = -\frac{\mu}{E}, \quad s_{44} = \frac{2\left(1+\mu\right)}{E}, \tag{16}$$

The Young' modulus E and the Poisson's ratio μ are related to the cylindrical particle ($q = p$) and the cell matrix ($q = m$). Consequently, the transformations $E \to E_q$, $\mu \to \mu_q$ ($i, j = 1, \ldots, 6$; $q = p,m$) are required to be considered.

4. Mathematical Procedures

4.1. Method of Separation of Variables

Let the Cauchy's equations (see Eqs. (3)–(7)) be substituted to the Hooke's law for an isotropic solid elastic continuum (see Eqs. (11)–(16)). Consequently, the radial stress σ'_{11} acting along the axis x'_1 (see Fig. 2), the tangential stress σ'_{22} acting along the axis x'_2, the tangential stress σ'_{33} acting along the axis x'_3, the shear stress σ'_{12} acting along the axis x'_2 and the shear stress σ'_{13} acting along the axis x'_3 have the forms

$$\sigma'_{11} = (c_1 + c_2)\,\frac{\partial u'_1}{\partial r} - c_2\left(\frac{u'_1}{r} + \frac{\partial u'_3}{\partial x_3}\right), \tag{17}$$

$$\sigma'_{22} = (c_1 + c_2)\,\frac{u'_1}{r} - c_2\left(\frac{\partial u'_1}{\partial r} + \frac{\partial u'_3}{\partial x_3}\right), \tag{18}$$

$$\sigma'_{33} = (c_1 + c_2)\,\frac{\partial u'_3}{\partial x_3} - c_2\left(\frac{\partial u'_1}{\partial r} + \frac{u'_1}{r}\right), \tag{19}$$

$$\sigma'_{12} = \frac{1}{s_{44} r}\,\frac{\partial u'_1}{\partial \varphi}, \tag{20}$$

$$\sigma'_{13} = \frac{1}{2 s_{44}}\left(\frac{\partial u'_1}{\partial x_3} + \frac{\partial u'_3}{\partial r}\right), \tag{21}$$

where $\sigma'_{ij} = \sigma'_{ji}$ $(i, j = 1,2,3)$ [22]–[24], and the coefficients c_1, c_2 are derived as (see Eq. (16))

$$c_1 = \frac{s_{11}}{s_{11}\,(s_{11} + s_{12}) - 2 s_{12}^2} = \frac{E}{(1+\mu)\,(1-2\mu)}, \quad c_2 = \frac{s_{12}\, c_1}{s_{11}} = -\mu\, c_1. \tag{22}$$

Let the equations (17)–(21) be substituted to Eq. (8), the equilibrium equation (8) is thus transformed to the form

$$r^2\,\frac{\partial^2 u'_1}{\partial r^2} + r\,\frac{\partial u'_1}{\partial r} - u'_1 + \frac{1-2\mu}{2\,(1-\mu)}\left\{\frac{\partial^2 u'_1}{\partial \varphi^2} + \frac{r^2}{2}\left[\frac{\partial^2 u'_1}{\partial x_3^2} + \frac{1+2\mu}{1-2\mu}\,\frac{\partial^2 u'_1}{\partial r \partial x_3}\right]\right\} = 0. \tag{23}$$

Similarly, let the equations (17)–(21) be substituted to the $[\partial \mathrm{Eq.}(9)/\partial\varphi]$, the equilibrium equation (9) is thus transformed to the form

$$r\,\frac{\partial}{\partial r}\left(\frac{\partial^2 u'_1}{\partial \varphi^2}\right) + (3-4\mu)\,\frac{\partial^2 u'_1}{\partial \varphi^2} = 0. \tag{24}$$

Finally, let the equations (17)–(21) be substituted to Eq. (10), the equilibrium equation (10) is thus transformed to the form

$$r\,\frac{\partial^2 u_3'}{\partial x_3^2} + \frac{1-2\mu}{4\,(1-\mu)}\left[r\,\frac{\partial^2 u_3'}{\partial r^2} + \frac{\partial u_3'}{\partial r} + \frac{1+2\mu}{1-2\mu}\left(r\,\frac{\partial^2 u_1'}{\partial r \partial x_3} + \frac{\partial u_1'}{\partial x_3}\right)\right] = 0. \tag{25}$$

4.2. Radial and Axial Displacements

With regard to the system of the partial differential equations (23), (24), (25), solutions for the displacements $u_1' = u_1'(r, \varphi, x_3)$ and $u_3' = u_3'(r, x_3)$ along the axes x_1' and x_3', respectively, is obtained by a method of separation of variables [25]. With regard to the method of separation of variables r, φ, x_3, the displacements $u_1' = u_1'(r, \varphi, x_3)$ and $u_3' = u_3'(r, x_3)$ are assumed to exhibit the forms

$$u_1' = u_1'(r, \varphi, x_3) = u_{11}'(r, \varphi) \times u_{13}'(x_3) \tag{26}$$

$$u_3' = u_3'(r, x_3) = u_{33}'(x_3) \times u_{31}'(r)\,. \tag{27}$$

Accordingly, $u_{11}' = u_{11}'(r, \varphi)$ and $u_{13}' = u_{13}'(x_3)$, both related to $u_1' = u_1'(r, \varphi, x_3)$, are functions of the variables r, φ and x_3, respectively. Similarly, $u_{31}' = u_{31}'(r)$ and $u_{33}' = u_{33}'(x_3)$, both related to $u_3' = u_3'(r, x_3)$, are functions of the variables r, and x_3, respectively.

With regard to Eqs. (26), (27), the partial differential equation (23) is transformed to the following two partial differential equations derived as

$$r^2\,\frac{\partial^2 u_{11}'}{\partial r^2} + r\,\frac{\partial u_{11}'}{\partial r} - u_{11}' + \frac{1-2\mu}{2\,(1-\mu)}\,U_{11} = 0. \tag{28}$$

$$u_{11}'\,\frac{\partial^2 u_{13}'}{\partial x_3^2} + \frac{1+2\mu}{1-2\mu}\,\frac{\partial u_{13}'}{\partial r}\,\frac{\partial u_{33}'}{\partial x_3} = 0. \tag{29}$$

where the function U_{11} is derived as

$$U_{11} = \frac{\partial^2 u_{11}'}{\partial \varphi^2}. \tag{30}$$

Similarly, with regard to Eqs. (26), (27), (29), the partial differential equation (24) is transformed to the form

$$r\,\frac{\partial U_{11}}{\partial r} + (3-4\mu)\,U_{11} = 0, \tag{31}$$

Similarly, with regard to Eqs. (26), (27), the partial differential equation (25) is transformed to the form

$$\frac{1+2\mu}{2\,(1-2\mu)}\,\frac{\partial u_{13}'}{\partial x_3}\left[\frac{1}{r\,u_{31}'}\,\frac{\partial}{\partial r}\left(r\,u_{11}'\right)\right] + u_{33}'\left[\frac{1}{r\,u_{31}'}\,\frac{\partial}{\partial r}\left(r\,\frac{\partial u_{31}'}{\partial r}\right)\right] + \frac{2\,(1-\mu)}{1-2\mu}\,\frac{\partial^3 u_{33}'}{\partial x_3\,\partial x_3} = 0. \tag{32}$$

4.2.1. Function $u_{11q}' = u_{11q}'\,(r,\varphi)$

Performing $r\,[\partial\mathrm{Eq.}(24)/\partial r]$, we get

$$r^2\,\frac{\partial^2 U_{11}}{\partial r^2} + 4\,(1-\mu)\,r\,\frac{\partial U_{11}}{\partial r} = 0. \tag{33}$$

Substituting Eq. (24) to Eq. (33), we get

$$r^2\,\frac{\partial^2 U_{11}}{\partial r^2} - (3-4\mu)\,4\,(1-\mu)\,U_{11} = 0. \tag{34}$$

If the function $U_{11} = U_{11}\,(r)$ is assumed in the form $U_{11} = r^{\lambda}$, then the solution U_{11} of the differential equation (34) has the form

$$U_{11} = \frac{\partial^2 u_{11}'}{\partial \varphi^2} = \sum_{i=1}^{2} C_i^{(11)}\, r^{\lambda_i^{(11)}}, \tag{35}$$

where the integration constant $C_i^{(11)}$ $(i=1{,}2)$ is determined by the boundary conditions (see Sec. (6.2)), and the exponent λ_i $(i=1{,}2)$ is derived as

$$\lambda_i^{(11)} = \frac{1}{2}\left[1 + (\delta_{1i}-\delta_{2i})\,\sqrt{D}\right], \quad D = 1 + 16\,(1-\mu)\,[1+4\,(1-\mu)]\,, \quad i = 1,2, \tag{36}$$

With regard to the Poisson's number of an elastic solid continuum and real materials, $\mu = 0.25$ [22] and $\mu \in (0, 0.5)$ [9], respectively, we get for the discriminant $D > 0$, and accordingly for the real exponents $\lambda_1^{(11)} > 3$, $\lambda_2^{(11)} < -2$ with respect to $\mu \in (0, 0.5)$.

Considering Eq. (35), the partial differential equation (28) is transformed to the form

$$\frac{\partial^2 u'_{11}}{\partial r^2} + \frac{1}{r}\frac{\partial u'_{11}}{\partial r} - \frac{u'_{11}}{r^2} = -\frac{1-2\mu}{2(1-\mu)}\sum_{i=1}^{2} C_i^{(11)} r^{\lambda_i^{(11)}-2}. \tag{37}$$

Using the Wronskian's method [25], the function $u'_{11q} = u'_{11q}(r, \varphi)$ (see Eq. (26)) which represents a solution of Eq. (28) for the cylindrical particle ($q = p$) and the cell matrix ($q = m$) has the form

$$u'_{11q} = \sum_{i=1}^{2} C_{iq}^{(11)} u_{iq}^{(11)}. \tag{38}$$

The integration constant $C_{iq}^{(11)} = C_{iq}^{(11)}(\varphi)$ ($i = 1,2$) as a function of the variable $\varphi \in \langle 0, \pi/2 \rangle$ is determined by the boundary conditions (see Eqs. (53), (58), (63)–(73)) for the cylindrical particle ($q = p$) and the cell matrix ($q = m$).

The two solutions $u_{1q}^{(11)} = u_{1q}^{(11)}(r)$, $u_{2q}^{(11)} = u_{2q}^{(11)}(r)$ of the partial differential equation (37), and consequently the coefficients $\xi_{1q}^{(11)}$, $\xi_{2q}^{(11)}$ ($q = p,m$) have the forms

$$u_{iq}^{(11)} = \xi_{iq}^{(11)} r^{\lambda_{iq}^{(11)}}, \quad i = 1, 2; \quad \lambda_{1q}^{(11)} > 3, \quad \lambda_{2q}^{(11)} < -2 \tag{39}$$

$$\xi_{iq}^{(11)} = -\frac{1-2\mu_q}{2(1-\mu_q)\left\{\left[\lambda_{iq}^{(11)}\right]^2 - 1\right\}}, \quad i = 1, 2, \tag{40}$$

where $u_{1q}^{(11)} = u_{1q}^{(11)}(r)$ and $u_{2q}^{(11)} = u_{2q}^{(11)}(r)$ are increasing and decreasing functions of the variable r due to $\lambda_{1q}^{(11)} > 3$ and $\lambda_{2q}^{(11)} < -2$, respectively.

4.2.2. Function $u'_{33q} = u'_{33q}(x_3)$

The partial differential equation (32) can be solved provided that the following two conditions are required to be both valid

$$\frac{1}{r\,u'_{31q}}\frac{\partial}{\partial r}\left(r\,u'_{11q}\right) = 1 \quad \wedge \quad \frac{1}{r\,u'_{31q}}\frac{\partial}{\partial r}\left(r\,\frac{\partial u'_{31q}}{\partial r}\right) = 1, \tag{41}$$

and consequently, these two conditions result in the following partial differential equation derived as

$$\frac{\partial u'_{31q}}{\partial r} = u'_{11q}. \tag{42}$$

With regard to Eq. (42), the partial differential equation (29) is transformed to the form

$$\frac{\partial^2 u'_{13q}}{\partial x_3^2} = -\frac{1+2\mu_q}{1-2\mu_q}\frac{\partial u'_{33q}}{\partial x_3}. \tag{43}$$

and then we get

$$\frac{\partial u'_{13q}}{\partial x_3} = -\frac{1+2\mu_q}{1-2\mu_q}u'_{33q}. \tag{44}$$

With regard to Eqs. (42), (44), the partial differential equation (32) is derived as

$$\frac{\partial^3 u'_{33q}}{\partial x_3\,\partial x_3} - \frac{2\mu_q}{1-2\mu_q}u'_{33q} = 0. \tag{45}$$

If the function $u'_{33q} = u'_{33q}(x_3)$ (see Eq. (27)) is assumed in the form $u'_{33q} = e^{\lambda_q x_3}$, then the solution $u'_{33q} = u'_{33q}(x_3)$ of the differential equation (32) has the form

$$u'_{33q} = C_{1q}^{(33)}\,e^{\lambda_q^{(33)}x_3} + C_{2q}^{(33)}\,e^{-\lambda_q^{(33)}x_3}, \tag{46}$$

where the integration constant $C_i^{(33)}$ $(i=1,2)$ which is not a function of the variables r, φ, x_3 (see Eq. (2)), $\varphi \in \langle 0, \pi/2\rangle$ (i.e. $C_{iq}^{(33)} \neq f(r,\varphi,x_3)$) is determined by the boundary conditions (see Eqs. (54)–(56), (74)–(69)) for the cylindrical particle $(q=p)$ and the cell matrix $(q=m)$. The exponent $\lambda_q^{(33)}$ is derived as

$$\lambda_q^{(33)} = \sqrt{\frac{2\mu_q}{1-2\mu_q}}, \tag{47}$$

where $(1-2\mu_q) > 0$, and then $[2\mu_q/(1-2\mu_q)] > 0$ due to $\mu_q \in (0, 0.5)$ for real materials [9].

4.2.3. Function $u'_{13q} = u'_{13q}(r, \varphi)$

Finally, with regard to Eqs. (44), (47), the function $u'_{13q} = u'_{13q}(r, \varphi)$ has the form

$$u'_{13q} = -\frac{1 + 2\mu_q}{\sqrt{2\mu_q(1 - 2\mu_q)}}\left[C^{(13)}_{1q}\, e^{\lambda^{(33)}_q x_3} - C^{(13)}_{2q}\, e^{-\lambda^{(33)}_q x_3}\right] + C^{(13)}_{3q}, \quad (48)$$

where the integration constant $C^{(13)}_{iq}$ ($i = 1,2,3$) which is not a function of the variables r, φ, x_3 (i.e. $C^{(13)}_{iq} \neq f(r, \varphi, x_3)$) is determined by the boundary conditions (see Eq. (62)) for the cylindrical particle ($q = p$) and the cell matrix ($q = m$).

4.2.4. Function $u'_{31q} = u'_{31q}(r)$

With regard to Eqs. (39), (42), the function $u'_{31q} = u'_{31q}(r)$ is derived as

$$u'_{31q} = \frac{\xi^{(11)}_{1q}\, C^{(31)}_{1q}}{\lambda^{(11)}_{1q} + 1}\, r^{\lambda^{(11)}_{1q}+1} + \frac{\xi^{(11)}_{2q}\, C^{(31)}_{2q}}{\lambda^{(11)}_{2q} + 1}\, r^{\lambda^{(11)}_{2q}+1} + C^{(31)}_{3q}, \quad (49)$$

where the integration constant $C^{(31)}_{iq}$ ($i = 1,2,3$) which is not a function of the variables r, φ, x_3 (i.e. $C^{(31)}_{iq} \neq f(r, \varphi, x_3)$) is determined by the boundary conditions (see Eq. (60)) for the cylindrical particle ($q = p$) and the cell matrix ($q = m$).

5. Reason of Thermal Stresses

The thermal stresses originate during a cooling process at the temperature $T \in \langle T_f, T_r \rangle$, where T_f and T_r [26] are final and relaxation temperature of the cooling process, respectively. As defined in [26], the relaxation temperature T_r is such temperature below that the stress relaxation as a consequence of thermal-activated processes does not occur in a material. The relaxation temperature is defined approximately by the relationship $T_r = (0.35 - 0.4) \times T_m$ [26] and exactly by an experiment, where T_m is a melting point of a two-component material.

If precipitates are formed from a liquid matrix of the two-component material, then the melting point T_m represents minimum of the set $\{T_{mp}, T_{mm}\}$, where T_{mp} and T_{mm} are melting points of the precipitate (particle) and the matrix [26], respectively. If the precipitates are formed from a solid matrix, then T_m represents a melting point of the two-component material [26].

The thermal stresses which originate at the temperature $T \in \langle T_f, T_r \rangle$ are thus a consequence of the condition $\beta_m - \beta_p \neq 0$. The coefficient β_q is derived as

$$\beta_q = \int_T^{T_r} \alpha_q \, dT, \quad q = p, m, \tag{50}$$

where α_q is a thermal expansion coefficient of the cylindrical particle ($q = p$) and the matrix ($q = m$).

Within the analytical modelling presented in this chapter, this cooling process is characterized by a homogeneous temperature change. Using the cylindrical coordinates (r, φ, x_3) (see Fig. 2), the homogeneous temperature change is then characterized by the condition $\partial T/\partial r = \partial T/\partial \varphi = \partial T/\partial x_3 = 0$.

6. Determination of Radial and Axial Displacements

6.1. Analysis of Radial and Axial Displacements

With regard to the boundary conditions for the cylindrical particle and the cell matrix, the following analysis of displacements which is based on a concept of imaginary separation is required to be considered.

As presented in Sec. 5, the thermal stresses originate at the temperature $T \in \langle T_f, T_r \rangle$. Let the cylindrical particles and the infinite matrix be imaginarily separated, and then cylindrical holes are periodically distributed in the infinite matrix at inter-hole distance (see Fig. 1). Let $T \in \langle T_f, T_r \rangle$ represent temperature of the separated cylindrical particles and of the infinite matrix with the cylindrical holes. If the temperature T increases or decreases within the interval $\langle T_f, T_r \rangle$, then the components which are imaginarily separated expand or contract, respectively. The expansion and contraction result in displacements of points in the components. Due to the imaginary separation, these displacements result from the temperature change, and not from the difference $\beta_m - \beta_p \neq 0$ (see Eq. (50)), where this difference is a reason of the thermal stresses.

In other words, let the cylindrical particles be embedded in the infinite matrix. Let $\Delta T = T - T_r \neq 0$ represent the temperature change. Let the condition $\beta_p = \beta_m$ be considered. If $\beta_p = \beta_m$, then the thermal stresses do not originate in the multi-particle-matrix system, and the infinitesimal cylindrical cap is thus shifted due to the temperature change ΔT only and not due to the difference $\beta_m - \beta_p \neq 0$ i.e. not due to $(\alpha_m - \alpha_p)\,\Delta T$ provided that α_q ($q = p,m$) is not a function of the temperature T. This shift (displacement) of the infinitesimal cylindrical cap is not induced by the thermal stresses.

Let R_r, L_r represent dimensions of both the cylindrical particle and the cylindrical hole at the temperature $T = T_r$. The radius R_q and the length L_q of both the cylindrical particle ($q = p$) and the cylindrical hole ($q = m$) at the temperature $T < T_r$ are derived as

$$R_q = R_q\,(1 - \beta_q)\,, \quad T < T_r, \tag{51}$$

$$L_q = L_r\,(1 - \beta_q)\,, \quad T < T_r, \tag{52}$$

where R_q is not a function of $x_3 \in \langle 0, L_q/2\rangle$ due to the fact that β_q is not a function of x_3 and L_q is not a function of $r \in \langle 0, R_q\rangle$ due to the fact that β_q is not a function of r, i.e. $R_q \neq f\,(x_3)$ and $L_q \neq f\,(r)$ due to $\beta_q \neq f\,(r, x_3)$, respectively. Additionally, let d_r represent inter-particle distance (= inter-hole distance) at the temperature $T = T_r$, and then we get $d = d_r\,(1 - \beta_m)$ at the temperature $T < T_r$.

The following analysis concerning the radial displacement u'_{11q} and the axial displacement u'_{33q} (q=p,m) (see Eqs. (26), (27)) explains a way how to put the separated cylindrical particle with the dimensions R_p, L_p into the cylindrical matrix hole with the dimensions R_m, L_m in a centre of the cubic cell on the condition $\beta_p < \beta_m$ which results in $R_p > R_m$, $L_p > L_m$, i.e. how to put the bigger cylindrical particle into the smaller cylindrical matrix hole in a centre of the cubic cell. This analysis is also valid for the condition $\beta_p > \beta_m$ which results in $R_p < R_m$. Additionally, due to the matrix infinity, the following analysis is valid for each cubic cell of the multi-particle-matrix system (see Fig. 1).

Let the condition $\beta_p < \beta_m$ be considered, and then we get $R_p > R_m$, $L_p > L_m$. To put a bigger cylindrical particle into a smaller cylindrical matrix hole in a centre of the cubic cell, the radii R_p, R_m are both required to be changed to R, as well as L_p, L_m are both required to be changed to L. The radial displacement $\left|\left(u'_{1p}\right)_{r=R_p}\right| = R - R_p \neq f\,(x_3)$ and $(u'_{1m})_{r=R_m} = R - R_m \neq f\,(x_3)$ in the cylindrical particle and the cylindrical matrix hole, respectively, are not functions of x_3 due to $R_q \neq f\,(x_3)$ ($q = p,m$). With regard to the Cartesian system $(Ox'_1x'_2x'_3)$ as well as from a mathematical point of view, the radial displacement u'_{1p} is considered to be negative, i.e. $u'_{1p} = -\left|u'_{1p}\right|$, and accordingly, the absolute value is considered. Similarly, the axial displacement $\left|\left(u'_{3q}\right)_{x_3=L_q/2}\right| = (L - L_q)/2 \neq f(r, \varphi)$ is not a function of both r and φ due to $L_q \neq f(r, \varphi)$ $(q = p,m)$. The same analysis is valid on the condition $\beta_p > \beta_m$.

Numerical values of R and L are determined by experimental methods for a real two-component material of the precipitate-matrix type with cylindrical precipitates. Due to $\beta_q \ll 1$ for $\alpha \propto 10^{-6}$ and $(T_r - T) \propto 10 - 10^2$ (see Eq. (50)) [26], the radius R_r and L_r at $T_r \propto 10^3$ are replaced by R and L, respectively. Accordingly, we get

$$\left(u'_{1m}\right)_{r=R} - \left(u'_{1p}\right)_{r=R} = R\left(\beta_m - \beta_p\right), \tag{53}$$

$$\left(u'_{3m}\right)_{x_3=L/2} - \left(u'_{3p}\right)_{x_3=L/2} = \frac{L}{2}\left(\beta_m - \beta_p\right). \tag{54}$$

The condition (53) represent one of the boundary conditions (see Sec. 6.2) for the determination of the integration constants $C_{1p}^{(1)}$, $C_{1m}^{(1)}$, $C_{2m}^{(1)}$ in Eqs. (57), (58), (61), (62). Similarly, the condition (54) represent one of the boundary conditions (see Sec. 6.2) for the determination of the integration constants $C_{1p}^{(3)}$, $C_{2p}^{(3)}$, $C_{1m}^{(3)}$, $C_{2m}^{(3)}$ in Eqs. (55), (56), (59), (60).

With regard to $\beta_p < \beta_m$, the matrix cylindrical part with dimensions R and $(d - L)/2$ which is defined within the intervals $r \in \langle 0, R\rangle$ and $x_3 \in \langle L/2, d/2\rangle$ is pushed along the axis $+x'_3$ by the cylindrical particle (see Fig. 2). On the contrary, the cylindrical particle is pushed by this matrix cylindrical part along the axis $-x'_3$ (see Fig. 2), i.e. along a direction which is opposite to the direction of the pushing of this the matrix cylindrical part. This pushing is homogeneous over the surface S_3 of the particle-matrix boundary with the normal x'_3 and with the surface area $A_3 = \pi R^2$, and then the axial displacement $u'_{3p} = u'_{3p}(x_3)$ as a function of $x_3 \in \langle 0, L/2\rangle$ and the axial displacement $u'_{3m} = u'_{3m}(x_3)$ as a function of $x_3 \in \langle L/2, d/2\rangle$ are not functions of both r and φ, i.e. $u'_{3p} = u'_{3p}(x_3) \neq f(r, \varphi)$ for $x_3 \in \langle 0, L/2\rangle$ and $u'_{3m} = u'_{3m}(x_3) \neq f(r, \varphi)$ for $x_3 \in \langle L/2, d/2\rangle$.

Accordingly, the absolute value $\left|u'_{3p}\right| = f(x_3) \neq f(r, \varphi)$ for $r \in \langle 0, R\rangle$, $\varphi \in \langle 0, \pi/2\rangle$, $x_3 \in \langle 0, L/2\rangle$ is required to represent an increasing function of x_3, exhibiting a maximum value on the particle-matrix boundary, i.e. for $x_3 = L/2$. Consequently, the absolute value $\left|u'_{3m}\right| = f(x_3) \neq f(r, \varphi)$ for $r \in \langle 0, R\rangle$, $\varphi \in \langle 0, \pi/2\rangle$, $x_3 \in \langle L/2, d/2\rangle$ is required to represent a decreasing function of x_3, exhibiting a maximum value on the particle-matrix boundary, i.e. for $x_3 = L/2$.

With regard to Eqs. (27), (46), (49), the axial displacement $u'_{3p} = u'_{3p}(x_3) \neq f(r,\varphi)$ in the cylindrical particle for $r \in \langle 0, R\rangle$, $\varphi \in \langle 0, \pi/2\rangle$, $x_3 \in \langle 0, L/2\rangle$ and the axial displacement $u'_{3m} = u'_{3m}(x_3) \neq f(r,\varphi)$ in the cell matrix for $r \in \langle 0, R\rangle$, $\varphi \in \langle 0, \pi/2\rangle$, $x_3 \in \langle L/2, d/2\rangle$ have the forms

$$u'_{3p} = u'_{3p}(x_3) = C^{(3)}_{1p}\, e^{\lambda^{(33)}_p x_3} + C^{(3)}_{2p}\, e^{-\lambda^{(33)}_p x_3},$$
$$r \in \langle 0, R\rangle, \quad \varphi \in \left\langle 0, \frac{\pi}{2}\right\rangle, \quad x_3 \in \left\langle 0, \frac{L}{2}\right\rangle, \tag{55}$$

$$u'_{3m} = u'_{3m}(x_3) = C^{(3)}_{1m}\, e^{\lambda^{(33)}_m x_3} + C^{(3)}_{2m}\, e^{-\lambda^{(33)}_m x_3},$$
$$r \in \langle 0, R\rangle, \quad \varphi \in \left\langle 0, \frac{\pi}{2}\right\rangle, \quad x_3 \in \left\langle \frac{L}{2}, \frac{d}{2}\right\rangle, \tag{56}$$

where $C^{(31)}_{1q} = C^{(31)}_{2q} = 0$, $C^{(33)}_{3q} \neq 0$ (see Eq. (49)), and $C^{(3)}_{iq} = C^{(33)}_{iq} C^{(33)}_{3q}$ $(q = p,m;\ i = 1,2)$.

Similarly, this pushing is also homogeneous over the surface S_1 of the particle-matrix boundary with the normal x'_1 and with the surface area $S_1 = 2\pi RL/2$, and then the radial displacement $u'_{1p} = u'_{1p}(r,\varphi)$ as a function of the variables $r \in \langle 0, R\rangle$, $\varphi \in \langle 0, \pi/2\rangle$ along with the radial displacement $u'_{1m} = u'_{1m}(r)$ as a function of the variables $r \in \langle R, r_c\rangle$, $\varphi \in \langle 0, \pi/2\rangle$ (see Fig. 3, Eq. (2)) are not functions of x_3 for $x_3 \in \langle 0, L/2\rangle$, i.e. $u'_{1p} = u'_{1p}(r,\varphi) \neq f(x_3)$ for $r \in \langle 0, R\rangle$, $\varphi \in \langle 0, \pi/2\rangle$, $x_3 \in \langle 0, L/2\rangle$ and $u'_{1m} = u'_{1m}(r,\varphi) \neq f(x_3)$ for $r \in \langle R, r_c\rangle$, $\varphi \in \langle 0, \pi/2\rangle$, $x_3 \in \langle 0, L/2\rangle$.

The conditions (53), (54) and the analysis mentioned above are also valid on the condition $\beta_p > \beta_m$. With regard to $\beta_p > \beta_m$, the matrix cylindrical part which is defined within the intervals $r \in \langle 0, R\rangle$ and $x_3 \in \langle L/2, d/2\rangle$ is pulled along the axis $+x'_3$ by the cylindrical particle (see Fig. 2). On the contrary, the cylindrical particle is pulled by this matrix cylindrical part along the axis $-x'_3$ (see Fig. 2). This pulling is also homogeneous over the surface S_3 of the particle-matrix boundary with the normal x'_3.

Accordingly, the absolute value $\left|u'_{1p}\right| = f(r,\varphi) \neq f(x_3)$ for $r \in \langle 0, R\rangle$, $\varphi \in \langle 0, \pi/2\rangle$, $x_3 \in \langle 0, L/2\rangle$ is required to represent an increasing function of r, exhibiting a maximum value on the particle-matrix boundary, i.e. for $r = R$. Consequently, the absolute value $|u'_{1m}| = f(r,\varphi) \neq f(x_3)$ for $r \in \langle R, r_c\rangle$,

$\varphi \in \langle 0, \pi/2 \rangle$, $x_3 \in \langle 0, L/2 \rangle$ is required to represent a decreasing function of r, exhibiting a maximum value on the particle-matrix boundary, i.e. for $r = R$. With regard to Eqs. (26), (38), (48), the radial displacement $u'_{1p} = u'_{1p}(r, \varphi) \neq f(x_3)$ in the cylindrical particle for $r \in \langle 0, R \rangle$, $\varphi \in \langle 0, \pi/2 \rangle$, $x_3 \in \langle 0, L/2 \rangle$ and the radial displacement $u'_{1m} = u'_{1m}(r, \varphi) \neq f(x_3)$ in the cell matrix for $r \in \langle R, r_c \rangle$, $\varphi \in \langle 0, \pi/2 \rangle$, $x_3 \in \langle 0, L/2 \rangle$ are derived as

$$u'_{1p} = u'_{1p}(r, \varphi) = C^{(1)}_{1p} \xi^{(11)}_{1p} r^{\lambda^{(11)}_{1p}}, \quad r \in \langle 0, R \rangle, \quad \varphi \in \left\langle 0, \frac{\pi}{2} \right\rangle, \quad x_3 \in \left\langle 0, \frac{L}{2} \right\rangle, \tag{57}$$

$$\begin{aligned} u'_{1m} &= u'_{1m}(r, \varphi) = C^{(1)}_{1m} \xi^{(11)}_{1m} r^{\lambda^{(11)}_{1m}} + C^{(1)}_{2m} \xi^{(11)}_{2m} r^{\lambda^{(11)}_{2m}}, \\ r &\in \langle R, r_c \rangle, \quad \varphi \in \left\langle 0, \frac{\pi}{2} \right\rangle, \quad x_3 \in \left\langle 0, \frac{L}{2} \right\rangle, \end{aligned} \tag{58}$$

where $C^{(33)}_{1q} = C^{(33)}_{2q} = 0$, $C^{(33)}_{3q} \neq 0$ (see Eq. (48)), and $C^{(1)}_{1p} = C^{(1)}_{1p}(\varphi) = C^{(33)}_{3p} C^{(11)}_{1p}(\varphi)$, $C^{(1)}_{im} = C^{(33)}_{3m} C^{(11)}_{im}(\varphi)$ for i=1,2. If $C^{(11)}_{2p} \neq 0$ then $(u'_{1m})_{r \to 0} \to \pm\infty$ due to $\lambda^{(11)}_{2p} < -2$ (see Eqs. (36), (39)), and accordingly the coefficient $C^{(11)}_{2p}$ is required to be $C^{(11)}_{2p} = 0$.

With regard to the axial displacement $u'_{3m} = u'_{33m}(x_3) \times u'_{31m}(r)$ (see Eq. (27)) in the cell matrix for $r \in \langle R, r_c \rangle$ (see Fig. 3, Eq. (2)), the absolute value $|u'_{3m}|$ is required to represent a decreasing function of r. Accordingly, the function $u'_{31m} = u'_{31m}(r)$ (see Eq. (49)) is required to represent a decreasing function of $r \in \langle R, r_c \rangle$ (see Fig. 3, Eq. (2)), exhibiting a maximum value on the particle-matrix boundary, i.e. for $r = R$. With regard to Eq. (49), we get $C^{(31)}_{1m} = 0$ due to $\lambda^{(11)}_{1m} > 3$ (see Eqs. (36), (39)), $C^{(31)}_{2m} \neq 0$ and $C^{(31)}_{3m} = 0$.

Additionally, with regard to the interval $x_3 \in \langle 0, L/2 \rangle$, the axial displacement $u'_{3m} = u'_{3m}(r, x_3)$ is required to fulfil the condition $(u'_{3m})_{r=R} = u'_{3p}(x_3) \neq f(r, \varphi)$ for $\varphi \in \langle 0, \pi/2 \rangle$ (see Eq. (55)). With regard to the interval $x_3 \in \langle L/2, d/2 \rangle$, the axial displacement $u'_{3m} = u'_{3m}(r, x_3)$ for $r = R$ (i.e. $(u'_{3m})_{r=R}$) is required to be equal to the axial displacement $u'_{3m} = u'_{3m}(x_3) \neq f(r, \varphi)$ given by Eq. (56) which is defined for the intervals $r \in \langle 0, R \rangle$, $\varphi \in \langle 0, \pi/2 \rangle$, $x_3 \in \langle 0, L/2 \rangle$.

With regard to these boundary conditions, the axial displacement $u'_{3m} = u'_{3m}(r, x_3)$ defined in the intervals $x_3 \in \langle 0, L/2 \rangle$ and $x_3 \in \langle L/2, d/2 \rangle$, both for $r \in \langle R, r_c \rangle$ and $\varphi \in \langle 0, \pi/2 \rangle$ (see Fig. 3, Eq. (2)), has the forms

$$u'_{3m} = u'_{3m}(r, x_3) = \left[C_{1p}^{(3)} e^{\lambda_p^{(33)} x_3} + C_{2p}^{(3)} e^{-\lambda_p^{(33)} x_3} \right] \left(\frac{r}{R} \right)^{\lambda_{2m}^{(11)}},$$
$$r \in \langle R, r_c \rangle, \quad \varphi \in \left\langle 0, \frac{\pi}{2} \right\rangle, \quad x_3 \in \left\langle 0, \frac{L}{2} \right\rangle, \tag{59}$$

$$u'_{3m} = u'_{3m}(r, x_3) = \left[C_{1m}^{(3)} e^{\lambda_m^{(33)} x_3} + C_{2m}^{(3)} e^{-\lambda_m^{(33)} x_3} \right] \left(\frac{r}{R} \right)^{\lambda_{2m}^{(11)}},$$
$$r \in \langle R, r_c \rangle, \quad \varphi \in \left\langle 0, \frac{\pi}{2} \right\rangle, \quad x_3 \in \left\langle \frac{L}{2}, \frac{d}{2} \right\rangle, \tag{60}$$

where $C_{2m}^{(31)} = \left[\lambda_{2m}^{(11)} + 1 \right] / \left[\xi_{2m}^{(11)} R^{\lambda_{2m}^{(11)}} \right]$ for both intervals $x_3 \in \langle 0, L/2 \rangle$ and $x_3 \in \langle L/2, d/2 \rangle$.

With regard to the radial displacement $u'_{1m} = u'_{11m}(r, \varphi) \times u'_{31m}(x_3)$ (see Eq. (26)) in the cell matrix for $\varphi \in \langle 0, \pi/2 \rangle$, $x_3 \in \langle L/2, d/2 \rangle$, the absolute value $|u'_{1m}|$ is required to represent a decreasing function of x_3. Accordingly, the function $u'_{13m} = u'_{13m}(x_3)$ (see Eq. (48)) is required to represent a decreasing function of $x_3 \in \langle L/2, d/2 \rangle$, exhibiting a maximum value on the particle-matrix boundary, i.e. for $x_3 = L/2$. With regard to Eq. (48), we get $C_{1m}^{(13)} = 0$ due to $\lambda_m^{(33)} > 3$ (see (see Eq. (47))), $C_{2m}^{(13)} \neq 0$ and $C_{3m}^{(13)} = 0$.

Additionally, with regard to the interval $r \in \langle 0, R \rangle$, the radial displacement $u'_{1m} = u'_{1m}(r, \varphi, x_3)$ is required to fulfil the condition $(u'_{1m})_{x_3 = L/2} = u'_{1p}(r, \varphi) \neq f(x_3)$ (see Eq. (57)). With regard to the interval $r \in \langle R, r_c \rangle$ (see Fig. 3, Eq. (2)), the radial displacement $u'_{1m} = u'_{1m}(r, \varphi, x_3)$ for $x_3 = L/2$ (i.e. $(u'_{1m})_{x_3 = L/2}$) is required to be equal to the radial displacement $u'_{1m} = u'_{1m}(r, \varphi) \neq f(x_3)$ given by Eq. (58) which defined for the intervals $r \in \langle R, r_c \rangle$ and $x_3 \in \langle 0, L/2 \rangle$. With regard to these boundary conditions, the radial displacement $u'_{1m} = u'_{1m}(r, \varphi, x_3)$ defined in the intervals $r \in \langle 0, R \rangle$ and $r \in \langle R, r_c \rangle$ (see Fig. 3, Eq. (2)), both for $\varphi \in \langle 0, \pi/2 \rangle$, $x_3 \in \langle L/2, d/2 \rangle$, has the forms

$$u'_{1m} = u'_{1m}(r, \varphi, x_3) = C_{1p}^{(1)} \xi_{1p}^{(11)} r^{\lambda_{1p}^{(11)}} e^{-\lambda_m^{(33)} [x_3 - (L/2)]},$$
$$r \in \langle 0, R \rangle, \quad \varphi \in \left\langle 0, \frac{\pi}{2} \right\rangle, \quad x_3 \in \left\langle \frac{L}{2}, \frac{d}{2} \right\rangle, \tag{61}$$

$$u'_{1m} = u'_{1m}(r, \varphi, x_3) = \left[C_{1m}^{(1)} \xi_{1m}^{(11)} r^{\lambda_{1m}^{(11)}} + C_{2m}^{(1)} \xi_{2m}^{(11)} r^{\lambda_{2m}^{(11)}} \right] e^{-\lambda_m^{(33)} [x_3 - (L/2)]},$$

$$r \in \langle R, r_c \rangle, \quad \varphi \in \left\langle 0, \frac{\pi}{2} \right\rangle, \quad x_3 \in \left\langle \frac{L}{2}, \frac{d}{2} \right\rangle, \tag{62}$$

where $C_{2m}^{(13)} = e^{\lambda_m^{(33)} L/2} \left[\sqrt{2\mu_m (1 - 2\mu_m)} / (1 + 2\mu_m) \right]$ for both intervals $x_3 \in \langle 0, L/2 \rangle$ and $x_3 \in \langle L/2, d/2 \rangle$.

The integration constants $C_{1p}^{(1)}$, $C_{im}^{(1)}$, $C_{iq}^{(3)}$ ($i = 1,2$; $q = p,m$) in Eqs. (55)–(62) are determined by the boundary conditions in Sec. 6.2.

6.2. Boundary Conditions

The boundary conditions for the determination of the integration constants $C_{1p}^{(1)}$, $C_{1m}^{(1)}$, $C_{2m}^{(1)}$ in Eqs. (57), (58) have the forms

$$\left(\sigma'_{11p} \right)_{r=R} = -p_1, \tag{63}$$

$$\left(\sigma'_{11p} \right)_{r=R} = -p_1, \tag{64}$$

$$\left(u'_{1m} \right)_{r=r_c} = 0, \tag{65}$$

and additionally, the boundary condition (53) is also considered for the determination of the radial stress p_1 acting at the particle-matrix boundary for $r = R$, $\varphi \in \langle 0, \pi/2 \rangle$, $x_3 \in \langle 0, L/2 \rangle$.

With regard to Eqs. (17), (57), (58), (63)–(65), we get

$$C_{im}^{(1)} = -\frac{P_1}{\xi_{im}^{(11)} \xi_{2+im}^{(11)} \xi_{4+im}^{(11)} R^{\lambda_{im}^{(11)} - 1}}, \quad i = 1, 2, \tag{66}$$

$$C_{1p}^{(1)} = -\frac{P_1}{\xi_{1p}^{(11)} \xi_{3p}^{(11)} R^{\lambda_{1p}^{(11)} - 1}}, \tag{67}$$

where the coefficient $\xi_{iq}^{(11)}$ ($i = 1,2$; $q = p,m$) is given by Eq. (40), and the coefficients $\xi_{2+iq}^{(11)}$, $\xi_{4+im}^{(11)}$ are derived as

$$\xi_{2+iq}^{(11)} = \lambda_{iq}^{(11)} (c_{1q} + c_{2q}) - c_{2q}, \quad i = 1, 2; \quad q = p, m, \tag{68}$$

$$\xi_{4+im}^{(11)} = 1 - \frac{\xi_{5-im}^{(11)}}{\xi_{2+im}^{(11)}} \left(\frac{R}{r_c} \right)^{\lambda_{3-im}^{(11)} - \lambda_{im}^{(11)}}, \quad i = 1, 2. \tag{69}$$

The boundary conditions for the determination of the integration constants $C_{1p}^{(3)}$, $C_{2p}^{(3)}$ $C_{1m}^{(3)}$, $C_{2m}^{(3)}$ in Eqs. (55), (56) are derived as

$$\left(u'_{3p}\right)_{x_3=0} = 0, \tag{70}$$

$$\left(u'_{3m}\right)_{x_3=d/2} = 0, \tag{71}$$

$$\left(\frac{\partial^2 u'_{3p}}{\partial x_3^2}\right)_{x_3=L/2} = \left(\frac{\partial^2 u'_{3m}}{\partial x_3^2}\right)_{x_3=L/2}, \tag{72}$$

and additionally, the boundary condition (54) is also considered.

With regard to Eqs. (55), (56), the axial displacement $\varepsilon'_{33p} = \varepsilon'_{33p}(x_3)$ and $\varepsilon'_{33m} = \varepsilon'_{33m}(x_3)$, which are defined within the intervals $r \in \langle 0, R\rangle$, $\varphi \in \langle 0, \pi/2\rangle$, represents functions of $x_3 \in \langle 0, L/2\rangle$ and $x_3 \in \langle L/2, d/2\rangle$, respectively. A connection of these functions for $x_3 = L/2$ is assumed to be 'smooth'. Due to this assumption of the 'smooth' connection for $x_3 = L/2$, these dependences $\varepsilon'_{33p} - x_3$ and $\varepsilon'_{33m} - x_3$ for $r \in \langle 0, R\rangle$, $\varphi \in \langle 0, \pi/2\rangle$ are thus required not to mutually create a singular point for $x_3 = L/2$. To fulfil this non-singularity assumption, a tangent of $\varepsilon'_{33p} = \varepsilon'_{33p}(x_3)$ for $x_3 = L/2$ is equal of a tangent of $\varepsilon'_{33m} = \varepsilon'_{33m}(x_3)$ for $x_3 = L/2$, i.e. $\left(\partial\varepsilon'_{33p}/\partial x_3\right)_{x_3=L/2} = \left(\partial\varepsilon'_{33m}/\partial x_3\right)_{x_3=L/2}$. With regard to Eq. (5), the condition $\left(\partial\varepsilon'_{33p}/\partial x_3\right)_{x_3=L/2} = \left(\partial\varepsilon'_{33m}/\partial x_3\right)_{x_3=L/2}$ results in the boundary condition given by Eq. (72.

With regard to Eqs. (55), (56), (70)–(72), we get

$$C_{im}^{(3)} = \frac{L\left(\beta_m - \beta_p\right)}{2\,\xi_i^{(3)}}, \quad i = 1, 2, \tag{73}$$

$$C_{1p}^{(3)} = -C_{2p}^{(3)} = \frac{L\left(\beta_m - \beta_p\right)}{2\,\xi_3^{(3)}}, \tag{74}$$

where β_m, β_q are given by Eq (50), and the coefficients $\xi_1^{(3)}$, $\xi_2^{(3)}$, $\xi_3^{(3)}$ have the forms

$$\xi_1^{(3)} = \frac{\left[\lambda_p^{(33)}\right]^2 - \left[\lambda_m^{(33)}\right]^2}{\left[\lambda_p^{(33)}\right]^2}\left[1 - e^{\lambda_m^{(33)}(d-L)}\right] e^{\lambda_m^{(33)} L/2}, \tag{75}$$

$$\xi_2^{(3)} = \xi_1^{(3)} e^{-\lambda_m^{(33)} d}, \tag{76}$$

$$\xi_3^{(3)} = \frac{2\left\{\left[\lambda_p^{(33)}\right]^2 - \left[\lambda_m^{(33)}\right]^2\right\}}{\left[\lambda_m^{(33)}\right]^2} \sinh\left[\frac{L\,\lambda_p^{(33)}}{2}\right]. \tag{77}$$

The coefficient $P_1 = P_1(\varphi)$, and the radial stress $p_1 = p_1(\varphi)$ acting at the particle-matrix boundary for $r = R$, $\varphi \in \langle 0, \pi/2\rangle$, $x_3 \in \langle 0, L/2\rangle$ are derived as

$$P_1 = p_1 - c_{2m}\frac{\partial u'_{3p}}{\partial x_3} = \frac{\beta_m - \beta_p}{\xi_1^{(1)}} = p_1 - \frac{L\,c_{2m}\,\lambda_p^{(33)} \cosh\left[\lambda_p^{(33)} x_3\right](\beta_m - \beta_p)}{\xi_3^{(3)}}, \tag{78}$$

$$p_1 = (\beta_m - \beta_p)\left\{\frac{1}{\xi_1^{(1)}} + \frac{L\,c_{2m}\,\lambda_p^{(33)} \cosh\left[\lambda_p^{(33)} x_3\right]}{\xi_3^{(3)}}\right\}, \tag{79}$$

where the coefficient $\xi_1^{(1)}$ has the form

$$\xi_1^{(1)} = \frac{\xi_{1p}^{(11)}}{\xi_{3p}^{(11)}} - \left[\frac{\xi_{1m}^{(11)}}{\xi_{3m}^{(11)}\,\xi_{5m}^{(11)}} + \frac{\xi_{2m}^{(11)}}{\xi_{4m}^{(11)}\,\xi_{6m}^{(11)}}\right]. \tag{80}$$

The boundary conditions for the determination of the axial stress p_3 acting at the particle-matrix boundary for $x_3 = L/2$ within the intervals $\varphi \in \langle 0, \pi/2\rangle$, $r \in \langle 0, R\rangle$ are derived as

$$\left(\sigma'_{33p}\right)_{x_3 = L/2} = -p_3, \tag{81}$$

$$\left(\sigma'_{33m}\right)_{x_3 = L/2} = -p_3. \tag{82}$$

With regard to Eqs. (19), (55)–(57), (61), (71), (75)–(69), the axial stress p_3 acting at the particle-matrix boundary for $x_3 = L/2$ within the intervals $\varphi \in \langle 0, \pi/2\rangle$, $r \in \langle 0, R\rangle$ has the form

$$p_3 = \frac{L(\beta_m - \beta_p)\left[\lambda_p^{(33)}\,\lambda_m^{(33)}\right]^2}{2\left[\xi_{9p}^{(11)} - \xi_{pm}^{(11)}\right]\left\{\left[\lambda_p^{(33)}\right]^2 - \left[\lambda_m^{(33)}\right]^2\right\}} \times \left\{\frac{\xi_{pm}^{(11)}(c_{1p} + c_{2p})}{\lambda_p^{(33)}} \coth\left[\frac{L\,\lambda_p^{(33)}}{2}\right] + \frac{\xi_{9p}^{(11)}(c_{1m} + c_{2m})}{\lambda_m^{(33)}} \coth\left[\frac{(d-L)\,\lambda_m^{(33)}}{2}\right]\right\}, \tag{83}$$

where the coefficient $\xi_{pm}^{(11)}$ is derived as

$$\xi_{pm}^{(11)} = -c_{2m}\left[1 + \lambda_{1p}^{(11)}\right]. \tag{84}$$

6.3. Radial and Axial Displacements

6.3.1. Cylindrical Particle

The radial displacement $u'_{1p} = u'_{1p}(r, \varphi)$ (see Eq. (57)) and the axial displacement $u'_{3p} = u'_{3p}(x_3)$ (see Eq. (55)) in the cylindrical particle both for $r \in \langle 0, R\rangle$, $\varphi \in \langle 0, \pi/2\rangle$, $x_3 \in \langle 0, L/2\rangle$ are derived as

$$u'_{1p} = u'_{1p}(r, \varphi) = -\frac{r(\beta_m - \beta_p)}{\xi_1^{(1)}\xi_{3p}^{(11)}}\left(\frac{r}{R}\right)^{\lambda_{1p}^{(11)}-1},$$
$$r \in \langle 0, R\rangle, \quad \varphi \in \left\langle 0, \frac{\pi}{2}\right\rangle, \quad x_3 \in \left\langle 0, \frac{L}{2}\right\rangle, \tag{85}$$

$$u'_{3p} = u'_{3p}(x_3) = \frac{L(\beta_m - \beta_p)}{\xi_3^{(3)}} \sinh\left[\lambda_p^{(33)} x_3\right],$$
$$r \in \langle 0, R\rangle, \quad \varphi \in \left\langle 0, \frac{\pi}{2}\right\rangle, \quad x_3 \in \left\langle 0, \frac{L}{2}\right\rangle. \tag{86}$$

6.3.2. Cell Matrix

Intervals $r \in \langle R, r_c\rangle$**,** $\varphi \in \langle 0, \pi/2\rangle$**,** $x_3 \in \langle 0, L/2\rangle$**.** The radial displacement $u'_{1m} = u'_{1m}(r, \varphi)$ (see Eq. (58)) and the axial displacement $u'_{3m} = u'_{3m}(r, x_3)$ (see Eq. (59)) in the cell matrix have the forms

$$u'_{1m} = u'_{1m}(r, \varphi) = -\frac{r(\beta_m - \beta_p)}{\xi_1^{(1)}}\sum_{i=1}^{2}\frac{1}{\xi_{2+im}^{(11)}\xi_{4+im}^{(11)}}\left(\frac{r}{R}\right)^{\lambda_{im}^{(11)}-1},$$
$$r \in \langle R, r_c\rangle, \quad \varphi \in \left\langle 0, \frac{\pi}{2}\right\rangle, \quad x_3 \in \left\langle 0, \frac{L}{2}\right\rangle, \tag{87}$$

$$u'_{3m} = u'_{3m}(r, x_3) = \frac{L(\beta_m - \beta_p)}{\xi_3^{(3)}} \sinh\left[\lambda_p^{(33)} x_3\right]\left(\frac{r}{R}\right)^{\lambda_{2m}^{(11)}},$$
$$r \in \langle R, r_c\rangle, \quad \varphi \in \left\langle 0, \frac{\pi}{2}\right\rangle, \quad x_3 \in \left\langle 0, \frac{L}{2}\right\rangle. \tag{88}$$

Intervals $r \in \langle 0, R\rangle$**,** $\varphi \in \langle 0, \pi/2\rangle$**,** $x_3 \in \langle L/2, d/2\rangle$**.** The radial displacement $u'_{1m} = u'_{1m}(r, \varphi, x_3)$ (see Eq. (61)) and the axial displacement $u'_{3m} = u'_{3m}(x_3)$ (see Eq. (56)) in the cell matrix are derived as

$$u'_{1m} = u'_{1m}(r, \varphi, x_3) = -\frac{r(\beta_m - \beta_p)}{\xi_1^{(1)}\,\xi_{3p}^{(11)}}\left(\frac{r}{R}\right)^{\lambda_{1p}^{(11)}-1} \times e^{-\lambda_m^{(33)}[x_3-(L/2)]},$$

$$r \in \langle 0, R\rangle, \quad \varphi \in \left\langle 0, \frac{\pi}{2}\right\rangle, \quad x_3 \in \left\langle \frac{L}{2}, \frac{d}{2}\right\rangle, \tag{89}$$

$$u'_{3m} = u'_{3m}(x_3) = \frac{L(\beta_m - \beta_p)}{2}\left[\frac{e^{\lambda_m^{(33)}x_3}}{\xi_1^{(3)}} + \frac{e^{-\lambda_m^{(33)}x_3}}{\xi_2^{(3)}}\right],$$

$$r \in \langle 0, R\rangle, \quad \varphi \in \left\langle 0, \frac{\pi}{2}\right\rangle, \quad x_3 \in \left\langle \frac{L}{2}, \frac{d}{2}\right\rangle. \tag{90}$$

Intervals $r \in \langle R, r_c\rangle$**,** $\varphi \in \langle 0, \pi/2\rangle$**,** $x_3 \in \langle L/2, d/2\rangle$**.** The radial displacement $u'_{1m} = u'_{1m}(r, \varphi, x_3)$ (see Eq. (62)) and the axial displacement $u'_{3m} = u'_{3m}(r, x_3)$ (see Eq. (60)) in the cell matrix have the forms

$$u'_{1m} = u'_{1m}(r, \varphi, x_3) = -\frac{r(\beta_m - \beta_p)}{\xi_1^{(1)}}\sum_{i=1}^{2}\frac{1}{\xi_{2+im}^{(11)}\,\xi_{4+im}^{(11)}}\left(\frac{r}{R}\right)^{\lambda_{im}^{(11)}-1}$$
$$e^{-\lambda_m^{(33)}[x_3-(L/2)]},$$

$$r \in \langle R, r_c\rangle, \quad \varphi \in \left\langle 0, \frac{\pi}{2}\right\rangle, \quad x_3 \in \left\langle \frac{L}{2}, \frac{d}{2}\right\rangle, \tag{91}$$

$$u'_{3m} = u'_{3m}(r, x_3) = \frac{L(\beta_m - \beta_p)}{2}\left[\frac{e^{\lambda_m^{(33)}x_3}}{\xi_1^{(3)}} + \frac{e^{-\lambda_m^{(33)}x_3}}{\xi_2^{(3)}}\right]\left(\frac{r}{R}\right)^{\lambda_{2m}^{(11)}},$$

$$r \in \langle R, r_c\rangle, \quad \varphi \in \left\langle 0, \frac{\pi}{2}\right\rangle, \quad x_3 \in \left\langle \frac{L}{2}, \frac{d}{2}\right\rangle. \tag{92}$$

7. Thermal Strains and Stresses

7.1. Cylindrical Particle

With regard to Eqs. (3)–(7), (85), (86), the thermal strains in the cylindrical particle for $r \in \langle 0, R \rangle$, $\varphi \in \langle 0, \pi/2 \rangle$, $x_3 \in \langle 0, L/2 \rangle$ are derived as

$$\varepsilon'_{11p} = \varepsilon_{11p}(r,\varphi) = -\frac{(\beta_m - \beta_p)\,\lambda_{1p}^{(11)}}{\xi_1^{(1)}\,\xi_{3p}^{(11)}}\left(\frac{r}{R}\right)^{\lambda_{1p}^{(11)}-1},$$
$$r \in \langle 0, R \rangle, \quad \varphi \in \left\langle 0, \frac{\pi}{2} \right\rangle, \quad x_3 \in \left\langle 0, \frac{L}{2} \right\rangle, \tag{93}$$

$$\varepsilon'_{22p} = \varepsilon_{22p}(r,\varphi) = -\frac{(\beta_m - \beta_p)}{\xi_1^{(1)}\,\xi_{3p}^{(11)}}\left(\frac{r}{R}\right)^{\lambda_{1p}^{(11)}-1},$$
$$r \in \langle 0, R \rangle, \quad \varphi \in \left\langle 0, \frac{\pi}{2} \right\rangle, \quad x_3 \in \left\langle 0, \frac{L}{2} \right\rangle, \tag{94}$$

$$\varepsilon'_{33p} = \varepsilon_{33p}(x_3) = \frac{L\,\lambda_p^{(33)}\,(\beta_m - \beta_p)}{\xi_3^{(3)}}\cosh\left[\lambda_p^{(33)} x_3\right],$$
$$r \in \langle 0, R \rangle, \quad \varphi \in \left\langle 0, \frac{\pi}{2} \right\rangle, \quad x_3 \in \left\langle 0, \frac{L}{2} \right\rangle, \tag{95}$$

$$\varepsilon'_{12p} = \varepsilon'_{21p}(r,\varphi) = \frac{\beta_m - \beta_p}{\xi_{3p}^{(11)}\left[\xi_1^{(1)}\right]^2}\left[\frac{\partial \xi_1^{(1)}}{\partial \varphi}\right]\left(\frac{r}{R}\right)^{\lambda_{1p}^{(11)}-1},$$
$$r \in \langle 0, R \rangle, \quad \varphi \in \left\langle 0, \frac{\pi}{2} \right\rangle, \quad x_3 \in \left\langle 0, \frac{L}{2} \right\rangle, \tag{96}$$

where $\varepsilon'_{13} = \varepsilon'_{31} = 0$ due to $u'_{1p} = u'_{1p}(r,\varphi) \neq f(x_3)$, $u'_{3p} = u'_{3p}(x_3) \neq f(r)$ (see Eqs. (7), (85), (86)).

With regard to Eqs. (17)–(21), (85), (86), the thermal stresses in the cylindrical particle for $r \in \langle 0, R \rangle$, $\varphi \in \langle 0, \pi/2 \rangle$, $x_3 \in \langle 0, L/2 \rangle$ have the forms

$$\sigma'_{11p} = \sigma'_{11p}(r,\varphi,x_3) = -(\beta_m - \beta_p)\left\{\frac{1}{\xi_1^{(1)}}\left(\frac{r}{R}\right)^{\lambda_{1p}^{(11)}-1} + \frac{L c_{2p}\lambda_p^{(33)}}{\xi_3^{(3)}}\cosh\left[\lambda_p^{(33)} x_3\right]\right\},$$
$$r \in \langle 0, R \rangle, \quad \varphi \in \left\langle 0, \frac{\pi}{2} \right\rangle, \quad x_3 \in \left\langle 0, \frac{L}{2} \right\rangle, \tag{97}$$

$$\sigma'_{22p} = \sigma'_{22p}\,(r,\varphi,x_3) =$$
$$-\left(\beta_m - \beta_p\right)\left\{\frac{\xi_{7p}^{(11)}}{\xi_1^{(1)}\,\xi_{3p}^{(11)}}\left(\frac{r}{R}\right)^{\lambda_{1p}^{(11)}-1} + \frac{Lc_{2p}\lambda_p^{(33)}}{\xi_3^{(3)}}\,\cosh\left[\lambda_p^{(33)}x_3\right]\right\},$$
$$r \in \langle 0, R\rangle\,, \quad \varphi \in \left\langle 0, \frac{\pi}{2}\right\rangle, \quad x_3 \in \left\langle 0, \frac{L}{2}\right\rangle, \tag{98}$$

$$\sigma'_{33p} = \sigma'_{33p}\,(r,\varphi,x_3) =$$
$$-\left(\beta_m - \beta_p\right)\left\{\frac{\xi_{9p}^{(11)}}{\xi_1^{(1)}\,\xi_{3p}^{(11)}}\left(\frac{r}{R}\right)^{\lambda_{1p}^{(11)}-1} + \frac{L\left(c_{1p}+c_{2p}\right)\lambda_p^{(33)}}{\xi_3^{(3)}}\,\cosh\left[\lambda_p^{(33)}x_3\right]\right\},$$
$$r \in \langle 0, R\rangle\,, \quad \varphi \in \left\langle 0, \frac{\pi}{2}\right\rangle, \quad x_3 \in \left\langle 0, \frac{L}{2}\right\rangle, \tag{99}$$

$$\sigma'_{12p} = \sigma'_{21p}\,(r,\varphi) = \frac{\beta_m - \beta_p}{s_{44p}\,\xi_{3p}^{(11)}\left[\xi_1^{(1)}\right]^2}\left[\frac{\partial \xi_1^{(1)}}{\partial \varphi}\right]\left(\frac{r}{R}\right)^{\lambda_{1p}^{(11)}-1},$$
$$r \in \langle 0, R\rangle\,, \quad \varphi \in \left\langle 0, \frac{\pi}{2}\right\rangle, \quad x_3 \in \left\langle 0, \frac{L}{2}\right\rangle, \tag{100}$$

where $\sigma'_{13} = \sigma'_{31} = 0$ due to $u'_{1p} = u'_{1p}\,(r,\varphi) \neq f\,(x_3)$, $u'_{3p} = u'_{3p}\,(x_3) \neq f\,(r)$ (see Eqs. (21), (85), (86)). The coefficients $\xi_{7p}^{(11)}$, $\xi_{9p}^{(11)}$ are derived as

$$\xi_{6+iq}^{(11)} = c_{1q} + c_{2q}\left[1 - \lambda_{iq}^{(11)}\right], \quad i = 1,2, \quad q = p,m, \tag{101}$$

$$\xi_{8+iq}^{(11)} = -c_{2q}\left[1 + \lambda_{iq}^{(11)}\right], \quad i = 1,2, \quad q = p,m. \tag{102}$$

7.2. Cell Matrix

7.2.1. Intervals $r \in \langle R, r_c \rangle$, $\varphi \in \langle 0, \pi/2 \rangle$, $x_3 \in \langle 0, L/2 \rangle$

With regard to Eqs. (3)–(7), (87), (88), the thermal strains in the cell matrix for $r \in \langle 0, R \rangle$, $\varphi \in \langle 0, \pi/2 \rangle$, $x_3 \in \langle 0, L/2 \rangle$ have the forms

$$\varepsilon'_{11m} = \varepsilon'_{11m}(r, \varphi) = -\frac{\beta_m - \beta_p}{\xi_1^{(1)}} \sum_{i=1}^{2} \frac{\lambda_{im}^{(11)}}{\xi_{2+im}^{(11)} \xi_{4+im}^{(11)}} \left(\frac{r}{R}\right)^{\lambda_{im}^{(11)}-1},$$
$$r \in \langle R, r_c \rangle, \quad \varphi \in \left\langle 0, \frac{\pi}{2} \right\rangle, \quad x_3 \in \left\langle 0, \frac{L}{2} \right\rangle, \tag{103}$$

$$\varepsilon'_{22m} = \varepsilon'_{22m}(r, \varphi) = -\frac{\beta_m - \beta_p}{\xi_1^{(1)}} \sum_{i=1}^{2} \frac{1}{\xi_{2+im}^{(11)} \xi_{4+im}^{(11)}} \left(\frac{r}{R}\right)^{\lambda_{im}^{(11)}-1},$$
$$r \in \langle R, r_c \rangle, \quad \varphi \in \left\langle 0, \frac{\pi}{2} \right\rangle, \quad x_3 \in \left\langle 0, \frac{L}{2} \right\rangle, \tag{104}$$

$$\varepsilon'_{33m} = \varepsilon'_{33m}(r, x_3) = \frac{L(\beta_m - \beta_p)\lambda_p^{(33)}}{\xi_3^{(3)}} \cosh\left[\lambda_p^{(33)} x_3\right] \left(\frac{r}{R}\right)^{\lambda_{2m}^{(11)}},$$
$$r \in \langle R, r_c \rangle, \quad \varphi \in \left\langle 0, \frac{\pi}{2} \right\rangle, \quad x_3 \in \left\langle 0, \frac{L}{2} \right\rangle. \tag{105}$$

$$\varepsilon'_{12m} = \varepsilon'_{21m}(r, \varphi) = (\beta_m - \beta_p)$$
$$\times \sum_{i=1}^{2} \frac{1}{\xi_{2+im}^{(11)} \left[\xi_1^{(1)} \xi_{4+im}^{(11)}\right]^2} \left[\xi_{4+im}^{(11)} \frac{\partial \xi_1^{(1)}}{\partial \varphi} + \xi_1^{(1)} \frac{\partial \xi_{4+im}^{(11)}}{\partial \varphi}\right] \left(\frac{r}{R}\right)^{\lambda_{im}^{(11)}-1},$$
$$r \in \langle R, r_c \rangle, \quad \varphi \in \left\langle 0, \frac{\pi}{2} \right\rangle, \quad x_3 \in \left\langle 0, \frac{L}{2} \right\rangle, \tag{106}$$

$$\varepsilon'_{13m} = \varepsilon'_{31m}(r, x_3) = \frac{L(\beta_m - \beta_p)\lambda_{2m}^{(11)}}{R\,\xi_3^{(3)}} \sinh\left[\lambda_p^{(33)} x_3\right] \left(\frac{r}{R}\right)^{\lambda_{2m}^{(11)}-1},$$
$$r \in \langle R, r_c \rangle, \quad \varphi \in \left\langle 0, \frac{\pi}{2} \right\rangle, \quad x_3 \in \left\langle 0, \frac{L}{2} \right\rangle, \tag{107}$$

where the formula for the shear strain $\varepsilon'_{13} = \varepsilon'_{31}$ given by Eq. (7) is transformed to the form $\varepsilon'_{13} = \varepsilon'_{31} = \partial u'_{3m}/\partial r$ due to $u'_{1m} = u'_{1m}(r, \varphi) \neq f(x_3)$ (see Eqs. (87), (88)).

With regard to Eqs. (17)–(21), (87), (88), the thermal stresses in the cell matrix for $r \in \langle 0, R \rangle$, $\varphi \in \langle 0, \pi/2 \rangle$, $x_3 \in \langle 0, L/2 \rangle$ have the forms

$$\sigma'_{11m} = \sigma'_{11m}(r, \varphi) = -(\beta_m - \beta_p)$$
$$\times \left\{ \frac{1}{\xi_1^{(1)}} \sum_{i=1}^{2} \frac{1}{\xi_{4+im}^{(11)}} \left(\frac{r}{R} \right)^{\lambda_{im}^{(11)} - 1} + \frac{L\, c_{2m}\, \lambda_p^{(33)}}{\xi_3^{(3)}} \cosh\left[\lambda_p^{(33)} x_3 \right] \left(\frac{r}{R} \right)^{\lambda_{2m}^{(11)}} \right\},$$
$$r \in \langle R, r_c \rangle, \quad \varphi \in \left\langle 0, \frac{\pi}{2} \right\rangle, \quad x_3 \in \left\langle 0, \frac{L}{2} \right\rangle, \tag{108}$$

$$\sigma'_{22m} = \sigma'_{22m}(r, \varphi) = -(\beta_m - \beta_p)$$
$$\times \left\{ \frac{1}{\xi_1^{(1)}} \sum_{i=1}^{2} \frac{\xi_{6+im}^{(11)}}{\xi_{2+im}^{(11)}\, \xi_{4+im}^{(11)}} \left(\frac{r}{R} \right)^{\lambda_{im}^{(11)} - 1} + \frac{L\, c_{2m}\, \lambda_p^{(33)}}{\xi_3^{(3)}} \cosh\left[\lambda_p^{(33)} x_3 \right] \left(\frac{r}{R} \right)^{\lambda_{2m}^{(11)}} \right\},$$
$$r \in \langle R, r_c \rangle, \quad \varphi \in \left\langle 0, \frac{\pi}{2} \right\rangle, \quad x_3 \in \left\langle 0, \frac{L}{2} \right\rangle, \tag{109}$$

$$\sigma'_{33m} = \sigma'_{33m}(r, \varphi) = -(\beta_m - \beta_p) \times$$
$$\left\{ \frac{1}{\xi_1^{(1)}} \sum_{i=1}^{2} \frac{\xi_{8+im}^{(11)}}{\xi_{2+im}^{(11)}\, \xi_{4+im}^{(11)}} \left(\frac{r}{R} \right)^{\lambda_{im}^{(11)} - 1} + \frac{L\,(c_{1m} + c_{2m})\, \lambda_p^{(33)}}{\xi_3^{(3)}} \cosh\left[\lambda_p^{(33)} x_3 \right] \left(\frac{r}{R} \right)^{\lambda_{2m}^{(11)}} \right\},$$
$$r \in \langle R, r_c \rangle, \quad \varphi \in \left\langle 0, \frac{\pi}{2} \right\rangle, \quad x_3 \in \left\langle 0, \frac{L}{2} \right\rangle, \tag{110}$$

$$\sigma'_{12m} = \sigma'_{21m}(r, \varphi) = \frac{(\beta_m - \beta_p)}{s_{44m}}$$
$$\times \sum_{i=1}^{2} \frac{1}{\xi_{2+im}^{(11)} \left[\xi_1^{(1)}\, \xi_{4+im}^{(11)} \right]^2} \left[\xi_{4+im}^{(11)} \frac{\partial \xi_1^{(1)}}{\partial \varphi} + \xi_1^{(1)} \frac{\partial \xi_{4+im}^{(11)}}{\partial \varphi} \right] \left(\frac{r}{R} \right)^{\lambda_{im}^{(11)} - 1},$$
$$r \in \langle R, r_c \rangle, \quad \varphi \in \left\langle 0, \frac{\pi}{2} \right\rangle, \quad x_3 \in \left\langle 0, \frac{L}{2} \right\rangle, \tag{111}$$

$$\sigma'_{13m} = \sigma'_{31m}(r, x_3) = \frac{L(\beta_m - \beta_p)\,\lambda_{2m}^{(11)}}{R\,s_{44m}\,\xi_3^{(3)}} \sinh\left[\lambda_p^{(33)} x_3\right] \left(\frac{r}{R}\right)^{\lambda_{2m}^{(11)}-1},$$

$$r \in \langle R, r_c \rangle, \quad \varphi \in \left\langle 0, \frac{\pi}{2} \right\rangle, \quad x_3 \in \left\langle 0, \frac{L}{2} \right\rangle, \tag{112}$$

where the coefficients $\xi_{6+im}^{(11)}$, $\xi_{8+im}^{(11)}$ $(i = 1{,}2)$ are given by Eqs. (102), (103), and the formula for the shear stress $\sigma'_{13} = \sigma'_{31}$ given by Eq. (21) is transformed to the form $\sigma'_{13} = \sigma'_{31} = (1/s_{44m})\,(\partial u'_{3m}/\partial r)$ due to $u'_{1m} = u'_{1m}(r, \varphi) \neq f(x_3)$ (see Eqs. (87), (88)).

7.2.2. Intervals $r \in \langle 0, R \rangle$, $\varphi \in \langle 0, \pi/2 \rangle$, $x_3 \in \langle L/2, d/2 \rangle$

With regard to Eqs. (3)–(7), (89), (90), the thermal strains in the cell matrix for $r \in \langle 0, R \rangle$, $\varphi \in \langle 0, \pi/2 \rangle$, $x_3 \in \langle 0, L/2 \rangle$ are derived as

$$\varepsilon'_{11m} = \varepsilon'_{11m}(r, \varphi, x_3) = -\frac{(\beta_m - \beta_p)\,\lambda_{1p}^{(11)}}{\xi_1^{(1)}\,\xi_{3p}^{(11)}} \left(\frac{r}{R}\right)^{\lambda_{1p}^{(11)}-1} \times e^{-\lambda_m^{(33)}[x_3 - (L/2)]},$$

$$r \in \langle 0, R \rangle, \quad \varphi \in \left\langle 0, \frac{\pi}{2} \right\rangle, \quad x_3 \in \left\langle \frac{L}{2}, \frac{d}{2} \right\rangle, \tag{113}$$

$$\varepsilon'_{22m} = \varepsilon'_{22m}(r, \varphi, x_3) = -\frac{\beta_m - \beta_p}{\xi_1^{(1)}\,\xi_{3p}^{(11)}} \left(\frac{r}{R}\right)^{\lambda_{1p}^{(11)}-1} \times e^{-\lambda_m^{(33)}[x_3 - (L/2)]},$$

$$r \in \langle 0, R \rangle, \quad \varphi \in \left\langle 0, \frac{\pi}{2} \right\rangle, \quad x_3 \in \left\langle \frac{L}{2}, \frac{d}{2} \right\rangle, \tag{114}$$

$$\varepsilon'_{33m} = \varepsilon'_{33m}(x_3) = \frac{L(\beta_m - \beta_p)\,\lambda_m^{(33)}}{2} \left[\frac{e^{\lambda_m^{(33)} x_3}}{\xi_1^{(3)}} - \frac{e^{-\lambda_m^{(33)} x_3}}{\xi_2^{(3)}}\right],$$

$$r \in \langle 0, R \rangle, \quad \varphi \in \left\langle 0, \frac{\pi}{2} \right\rangle, \quad x_3 \in \left\langle \frac{L}{2}, \frac{d}{2} \right\rangle. \tag{115}$$

$$\varepsilon'_{12m} = \varepsilon'_{21m}(r, \varphi, x_3) = \frac{\beta_m - \beta_p}{\xi_{3p}^{(11)}\left[\xi_1^{(1)}\right]^2} \left[\frac{\partial \xi_1^{(1)}}{\partial \varphi}\right] \left(\frac{r}{R}\right)^{\lambda_{1p}^{(11)}-1} \times e^{-\lambda_m^{(33)}[x_3 - (L/2)]},$$

$$r \in \langle 0, R \rangle, \quad \varphi \in \left\langle 0, \frac{\pi}{2} \right\rangle, \quad x_3 \in \left\langle \frac{L}{2}, \frac{d}{2} \right\rangle, \tag{116}$$

$$\varepsilon'_{13m} = \varepsilon'_{31m}(r,\varphi,x_3) = \frac{r(\beta_m - \beta_p)\lambda_m^{(33)}}{\xi_1^{(1)}\xi_{3p}^{(11)}}\left(\frac{r}{R}\right)^{\lambda_{1p}^{(11)}-1} \times e^{-\lambda_m^{(33)}[x_3-(L/2)]},$$
$$r \in \langle 0, R\rangle, \quad \varphi \in \left\langle 0, \frac{\pi}{2}\right\rangle, \quad x_3 \in \left\langle \frac{L}{2}, \frac{d}{2}\right\rangle, \tag{117}$$

where the formula for the shear strain $\varepsilon'_{13} = \varepsilon'_{31}$ given by Eq. (7) is transformed to the form $\varepsilon'_{13} = \varepsilon'_{31} = \partial u'_{1m}/\partial x_3$ due to $u'_{3m} = u'_{3m}(x_3) \neq f(r)$ (see Eqs. (89), (90)).

With regard to Eqs. (17)–(21), (89), (90), the thermal stresses in the cell matrix for $r \in \langle 0, R\rangle$, $\varphi \in \langle 0, \pi/2\rangle$, $x_3 \in \langle 0, L/2\rangle$ have the forms

$$\sigma'_{11m} = \sigma'_{11m}(r,\varphi,x_3) = -(\beta_m - \beta_p)$$
$$\times \left\{ \frac{1}{\xi_1^{(1)}}\left(\frac{r}{R}\right)^{\lambda_{1p}^{(11)}-1} \times e^{-\lambda_m^{(33)}[x_3-(L/2)]} + \frac{Lc_{2m}\lambda_m^{(33)}}{2}\left[\frac{e^{\lambda_m^{(33)}x_3}}{\xi_1^{(3)}} - \frac{e^{-\lambda_m^{(33)}x_3}}{\xi_2^{(3)}}\right]\right\},$$
$$r \in \langle 0, R\rangle, \quad \varphi \in \left\langle 0, \frac{\pi}{2}\right\rangle, \quad x_3 \in \left\langle 0, \frac{L}{2}\right\rangle, \tag{118}$$

$$\sigma'_{22m} = \sigma'_{22m}(r,\varphi,x_3) = -(\beta_m - \beta_p)$$
$$\times \left\{ \frac{\xi_{7p}^{(11)}}{\xi_1^{(1)}\xi_{3p}^{(11)}}\left(\frac{r}{R}\right)^{\lambda_{1p}^{(11)}-1} \times e^{-\lambda_m^{(33)}[x_3-(L/2)]} + \frac{Lc_{2m}\lambda_m^{(33)}}{2}\left[\frac{e^{\lambda_m^{(33)}x_3}}{\xi_1^{(3)}} - \frac{e^{-\lambda_m^{(33)}x_3}}{\xi_2^{(3)}}\right]\right\},$$
$$r \in \langle 0, R\rangle, \quad \varphi \in \left\langle 0, \frac{\pi}{2}\right\rangle, \quad x_3 \in \left\langle 0, \frac{L}{2}\right\rangle, \tag{119}$$

$$\sigma'_{33m} = \sigma'_{33m}(r,\varphi,x_3) = -(\beta_m - \beta_p)\left\{ \frac{\xi_{9p}^{(11)}}{\xi_1^{(1)}\xi_{3p}^{(11)}}\left(\frac{r}{R}\right)^{\lambda_{1p}^{(11)}-1} \times e^{-\lambda_m^{(33)}[x_3-(L/2)]}\right.$$
$$\left. + \frac{L(c_{1m}+c_{2m})\lambda_m^{(33)}}{2}\left[\frac{e^{\lambda_m^{(33)}x_3}}{\xi_1^{(3)}} - \frac{e^{-\lambda_m^{(33)}x_3}}{\xi_2^{(3)}}\right]\right\},$$
$$r \in \langle 0, R\rangle, \quad \varphi \in \left\langle 0, \frac{\pi}{2}\right\rangle, \quad x_3 \in \left\langle 0, \frac{L}{2}\right\rangle, \tag{120}$$

$$\sigma'_{12m} = \sigma'_{21m}(r,\varphi,x_3) = \frac{\beta_m - \beta_p}{s_{44m}\,\xi_{3p}^{(11)}\left[\xi_1^{(1)}\right]^2}\left[\frac{\partial \xi_1^{(1)}}{\partial\varphi}\right]\left(\frac{r}{R}\right)^{\lambda_{1p}^{(11)}-1} \times e^{-\lambda_m^{(33)}[x_3-(L/2)]},$$
$$r \in \langle 0, R\rangle, \quad \varphi \in \left\langle 0, \frac{\pi}{2}\right\rangle, \quad x_3 \in \left\langle \frac{L}{2}, \frac{d}{2}\right\rangle, \tag{121}$$

$$\sigma'_{13m} = \sigma'_{31m}\left(r, \varphi, x_3\right) = \frac{r\left(\beta_m - \beta_p\right)\lambda_m^{(33)}}{s_{44m}\,\xi_1^{(1)}\,\xi_{3p}^{(11)}}\left(\frac{r}{R}\right)^{\lambda_{1p}^{(11)}-1} \times e^{-\lambda_m^{(33)}\left[x_3-(L/2)\right]},$$
$$r \in \langle 0, R\rangle, \quad \varphi \in \left\langle 0, \frac{\pi}{2}\right\rangle, \quad x_3 \in \left\langle \frac{L}{2}, \frac{d}{2}\right\rangle, \tag{122}$$

where the coefficients $\xi_{6+im}^{(11)}$, $\xi_{8+im}^{(11)}$ $(i = 1{,}2)$ are given by Eqs. (102), (103), and the formula for the shear stress $\varepsilon'_{13} = \varepsilon'_{31}$ given by Eq. (21) is transformed to the form $\sigma'_{13} = \sigma'_{31} = (1/s_{44m})\,\partial u'_{1m}/\partial x_3$ due to $u'_{3m} = u'_{3m}(x_3) \neq f(r)$ (see Eqs. (89), (90)).

7.2.3. Intervals $r \in \langle R, r_c\rangle$, $\varphi \in \langle 0, \pi/2\rangle$, $x_3 \in \langle L/2, d/2\rangle$

With regard to Eqs. (3)–(7), (91), (92), the thermal strains in the cell matrix for $r \in \langle 0, R\rangle$, $\varphi \in \langle 0, \pi/2\rangle$, $x_3 \in \langle 0, L/2\rangle$ are derived as

$$\varepsilon'_{11m} = \varepsilon'_{11m}\left(r, \varphi, x_3\right) = -\frac{\left(\beta_m - \beta_p\right)}{\xi_1^{(1)}}\sum_{i=1}^{2}\frac{\lambda_{im}^{(11)}}{\xi_{2+im}^{(11)}\,\xi_{4+im}^{(11)}}\left(\frac{r}{R}\right)^{\lambda_{im}^{(11)}-1} \times e^{-\lambda_m^{(33)}\left[x_3-(L/2)\right]},$$
$$r \in \langle R, r_c\rangle, \quad \varphi \in \left\langle 0, \frac{\pi}{2}\right\rangle, \quad x_3 \in \left\langle \frac{L}{2}, \frac{d}{2}\right\rangle, \tag{123}$$

$$\varepsilon'_{22m} = \varepsilon'_{22m}\left(r, \varphi, x_3\right) = -\frac{\beta_m - \beta_p}{\xi_1^{(1)}}\sum_{i=1}^{2}\frac{1}{\xi_{2+im}^{(11)}\,\xi_{4+im}^{(11)}}\left(\frac{r}{R}\right)^{\lambda_{im}^{(11)}-1} \times e^{-\lambda_m^{(33)}\left[x_3-(L/2)\right]},$$
$$r \in \langle R, r_c\rangle, \quad \varphi \in \left\langle 0, \frac{\pi}{2}\right\rangle, \quad x_3 \in \left\langle \frac{L}{2}, \frac{d}{2}\right\rangle, \tag{124}$$

$$\varepsilon'_{33m} = \varepsilon'_{33m}\left(r, x_3\right) = \frac{L\left(\beta_m - \beta_p\right)\lambda_m^{(33)}}{2}\left[\frac{e^{\lambda_m^{(33)}\,x_3}}{\xi_1^{(3)}} - \frac{e^{-\lambda_m^{(33)}\,x_3}}{\xi_2^{(3)}}\right]\left(\frac{r}{R}\right)^{\lambda_{2m}^{(11)}},$$
$$r \in \langle R, r_c\rangle, \quad \varphi \in \left\langle 0, \frac{\pi}{2}\right\rangle, \quad x_3 \in \left\langle \frac{L}{2}, \frac{d}{2}\right\rangle. \tag{125}$$

$$\varepsilon'_{12m} = \varepsilon'_{21m}\left(r, \varphi, x_3\right) = \left(\beta_m - \beta_p\right)$$
$$\times \sum_{i=1}^{2}\frac{1}{\xi_{2+im}^{(11)}\left[\xi_1^{(1)}\,\xi_{4+im}^{(11)}\right]^2}\left[\xi_{4+im}^{(11)}\frac{\partial \xi_1^{(1)}}{\partial \varphi} + \xi_1^{(1)}\frac{\partial \xi_{4+im}^{(11)}}{\partial \varphi}\right]\left(\frac{r}{R}\right)^{\lambda_{im}^{(11)}-1} \times e^{-\lambda_m^{(33)}\left[x_3-(L/2)\right]},$$
$$r \in \langle R, r_c\rangle, \quad \varphi \in \left\langle 0, \frac{\pi}{2}\right\rangle, \quad x_3 \in \left\langle \frac{L}{2}, \frac{d}{2}\right\rangle, \tag{126}$$

$$\varepsilon'_{13m} = \varepsilon'_{31m}(r,\varphi,x_3) = \frac{\beta_m - \beta_p}{2}\left\{\sum_{i=1}^{2} \frac{r\,\lambda_m^{(33)}}{\xi_1^{(1)}\,\xi_{2+im}^{(11)}\,\xi_{4+im}^{(11)}}\left(\frac{r}{R}\right)^{\lambda_{im}^{(11)}-1} \times e^{-\lambda_m^{(33)}[x_3-(L/2)]}\right.$$

$$\left. + \frac{L\,\lambda_{2m}^{(11)}}{2R}\left[\frac{e^{\lambda_m^{(33)}x_3}}{\xi_1^{(3)}} + \frac{e^{-\lambda_m^{(33)}x_3}}{\xi_2^{(3)}}\right]\left(\frac{r}{R}\right)^{\lambda_{2m}^{(11)}-1}\right\},$$

$$r \in \langle R, r_c\rangle,\quad \varphi \in \left\langle 0, \frac{\pi}{2}\right\rangle,\quad x_3 \in \left\langle \frac{L}{2}, \frac{d}{2}\right\rangle. \tag{127}$$

With regard to Eqs. (17)–(21), (91), (92), the thermal stresses in the cell matrix for $r \in \langle 0, R\rangle$, $\varphi \in \langle 0, \pi/2\rangle$, $x_3 \in \langle 0, L/2\rangle$ have the forms

$$\sigma'_{11m} = \sigma'_{11m}(r,\varphi) = -(\beta_m - \beta_p)\left\{\frac{1}{\xi_1^{(1)}}\sum_{i=1}^{2}\frac{1}{\xi_{4+im}^{(11)}}\left(\frac{r}{R}\right)^{\lambda_{im}^{(11)}-1} e^{-\lambda_m^{(33)}[x_3-(L/2)]}\right.$$

$$\left. + \frac{L\,c_{2m}\,\lambda_m^{(33)}}{2}\left[\frac{e^{\lambda_m^{(33)}x_3}}{\xi_1^{(3)}} - \frac{e^{-\lambda_m^{(33)}x_3}}{\xi_2^{(3)}}\right]\left(\frac{r}{R}\right)^{\lambda_{2m}^{(11)}}\right\},$$

$$r \in \langle R, r_c\rangle,\quad \varphi \in \left\langle 0, \frac{\pi}{2}\right\rangle,\quad x_3 \in \left\langle 0, \frac{L}{2}\right\rangle, \tag{128}$$

$$\sigma'_{22m} = \sigma'_{22m}(r,\varphi) = -(\beta_m - \beta_p)\left\{\frac{1}{\xi_1^{(1)}}\sum_{i=1}^{2}\frac{\xi_{6+im}^{(11)}}{\xi_{2+im}^{(11)}\,\xi_{4+im}^{(11)}}\left(\frac{r}{R}\right)^{\lambda_{im}^{(11)}-1} e^{-\lambda_m^{(33)}[x_3-(L/2)]}\right.$$

$$\left. + \frac{L\,c_{2m}\,\lambda_m^{(33)}}{2}\left[\frac{e^{\lambda_m^{(33)}x_3}}{\xi_1^{(3)}} - \frac{e^{-\lambda_m^{(33)}x_3}}{\xi_2^{(3)}}\right]\left(\frac{r}{R}\right)^{\lambda_{2m}^{(11)}}\right\},$$

$$r \in \langle R, r_c\rangle,\quad \varphi \in \left\langle 0, \frac{\pi}{2}\right\rangle,\quad x_3 \in \left\langle 0, \frac{L}{2}\right\rangle, \tag{129}$$

$$\sigma'_{33m} = \sigma'_{33m}(r,\varphi) = -(\beta_m - \beta_p)\left\{\frac{1}{\xi_1^{(1)}}\sum_{i=1}^{2}\frac{\xi_{8+im}^{(11)}}{\xi_{2+im}^{(11)}\,\xi_{4+im}^{(11)}}\left(\frac{r}{R}\right)^{\lambda_{im}^{(11)}-1} e^{-\lambda_m^{(33)}[x_3-(L/2)]}\right.$$

$$\left. + \frac{L\,(c_{1m}+c_{2m})\,\lambda_m^{(33)}}{2}\left[\frac{e^{\lambda_m^{(33)}x_3}}{\xi_1^{(3)}} - \frac{e^{-\lambda_m^{(33)}x_3}}{\xi_2^{(3)}}\right]\left(\frac{r}{R}\right)^{\lambda_{2m}^{(11)}}\right\},$$

$$r \in \langle R, r_c\rangle,\quad \varphi \in \left\langle 0, \frac{\pi}{2}\right\rangle,\quad x_3 \in \left\langle 0, \frac{L}{2}\right\rangle, \tag{130}$$

$$\sigma'_{12m} = \sigma'_{21m}(r,\varphi,x_3) = \frac{\beta_m - \beta_p}{s_{44m}}$$

$$\times \sum_{i=1}^{2}\frac{1}{\xi_{2+im}^{(11)}\left[\xi_1^{(1)}\,\xi_{4+im}^{(11)}\right]^2}\left[\xi_{4+im}^{(11)}\frac{\partial \xi_1^{(1)}}{\partial\varphi} + \xi_1^{(1)}\frac{\partial \xi_{4+im}^{(11)}}{\partial\varphi}\right]\left(\frac{r}{R}\right)^{\lambda_{im}^{(11)}-1} \times e^{-\lambda_m^{(33)}[x_3-(L/2)]},$$

$$r \in \langle R, r_c\rangle,\quad \varphi \in \left\langle 0, \frac{\pi}{2}\right\rangle,\quad x_3 \in \left\langle \frac{L}{2}, \frac{d}{2}\right\rangle, \tag{131}$$

$$\sigma'_{13m} = \sigma'_{31m}(r,\varphi,x_3) = \frac{\beta_m - \beta_p}{2s_{44m}} \left\{ \sum_{i=1}^{2} \frac{r\,\lambda_m^{(33)}}{\xi_1^{(1)}\,\xi_{2+im}^{(11)}\,\xi_{4+im}^{(11)}} \left(\frac{r}{R}\right)^{\lambda_{im}^{(11)}-1} \times e^{-\lambda_m^{(33)}[x_3-(L/2)]} \right.$$
$$\left. + \frac{L\,\lambda_{2m}^{(11)}}{2R} \left[\frac{e^{\lambda_m^{(33)} x_3}}{\xi_1^{(3)}} + \frac{e^{-\lambda_m^{(33)} x_3}}{\xi_2^{(3)}} \right] \left(\frac{r}{R}\right)^{\lambda_{2m}^{(11)}-1} \right\},$$
$$r \in \langle R, r_c \rangle, \quad \varphi \in \left\langle 0, \frac{\pi}{2} \right\rangle, \quad x_3 \in \left\langle \frac{L}{2}, \frac{d}{2} \right\rangle, \tag{132}$$

where the coefficients $\xi_{6+im}^{(11)}$, $\xi_{8+im}^{(11)}$ $(i = 1,2)$ are given by Eqs. (102), (103).

8. Application to SiC-Si_3N_4 Ceramics

This section presents illustrative examples of an application of this analytical model of thermal stresses in a two-component material with cylindrical particles to a real two-component material. These illustrative examples are numerically performed for the SiC-Si_3N_4 ceramics.

Table 1 presents material parameters of the SiC particle and the Si_3N_4 matrix of the SiC-Si_3N_4 ceramics representing a two-component material where the SiC particles can exhibit spherical and cylindrical shapes [26]. The relaxation temperature $T_r = 665$°C is determined by the formula $T_r = 0.35\,T_m$ [26], where $T_m = T_{mm}$ is melting temperature of the SiC-Si_3N_4 ceramics.

Table 1. Material parameters of the SiC-Si_3N_4 ceramics [26]

	Particle ($q = p$) **SiC**	**Matrix ($q = m$)** **Si_3N_4**
E_q [GPa]	360	310
μ_q	0.19	0.235
α_q [10^{-6}K^{-1}]	4.15	2.35
T_{mq} [°C]	2730	1900

As shown in Fig. 4, the dependences $\overline{p_1} - v$ and $p_3 - v$ (see Eqs. (1), (2), (79), (83)) for the radial stress $p_1 = p_1(\varphi, v)$ and the axial stress $p_3 = p_3(v)$ both acting at the SiC-Si_3N_4 boundary exhibit decreasing courses for $v \in (0, v_{pmax})$ (see Eq. (1)), where $L = 2R = 6$ μm, $T_r = 665$°C, $T_f = 20$°C. With regard

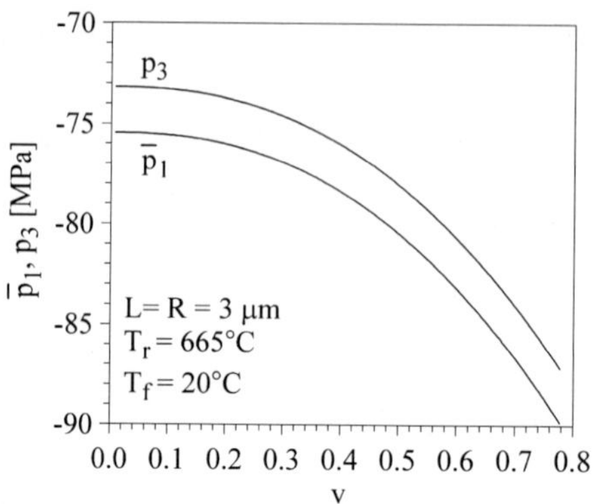

Figure 4. The functions $\overline{p_1} = \overline{p_1}(v)$ (see Eq. (133)) and $p_3 = p_3(v)$ of the volume fraction $v \in \langle 0, v_{pmax} \rangle$ (see Eq. (1)) of the SiC cylindrical particle, where $L = 2R = 6$ μm, $T_f = 20$°C, $T_r = 665$°C. The radial stress $p_1 = p_1(\varphi, v)$ and the axial stress $p_3 = p_3(v)$ (see Eqs. (79), (83)) act at the SiC-Si_3N_4 boundary.

to the interval $\varphi \in \langle 0, \pi/2 \rangle$, the mean value $\overline{p_1} = \overline{p_1}(v)$ of the function $p_1 = p_1(\varphi, v)$ is numerically determined by the following formula

$$\overline{p_1} = \frac{2}{\pi} \int_0^{\pi/2} p_1(\varphi)\, d\varphi \approx \sum_{i=0}^{n} p_1(i \times \Delta\varphi)\, \Delta\varphi, \tag{133}$$

where n is a integral part of the real number $\pi/(2\,\Delta\varphi)$. With regard to an experience of the author, the angle step $\Delta\varphi = 0.01 - 0.1$ [deg] is sufficient to be considered within the numerical determination of the mean vale.

Other applications of the final formulae (see Sec. 6) can be easily performed by a reader. This numerical determination can be performed by a computer software using suitable programming techniques. With regard to programming techniques within a suitable programming language (e.g. Fortran, Pascal), a derivative of the function $\psi = \psi(\varphi)$ regarding the variable φ (see Eqs. (96), (100), (106), (111), (116), (121), (126), (131)) can be numerically determined by the formula

$$\frac{\partial \psi}{\partial \varphi} \approx \frac{\psi(\varphi + \Delta\varphi) - \psi(\varphi)}{\Delta\varphi}, \tag{134}$$

where $\psi = \xi_1^{(1)}, \xi_{4+im}^{(11)}$ (see Eqs. (96), (100), (106), (111), (116), (121), (126),

(131)). The angle step $\Delta\varphi$ is required to be taken as small as possible. With regard to an experience of the author, the angle step $\Delta\varphi = 10^{-6}$ [deg] is sufficient to be considered within the numerical determination.

Conclusion

Results of this chapter are as follows.

The analytical model of thermal stresses in a multi-particle-matrix system with isotropic components is determined (see Secs. 2, 7). The analytical model results from fundamental equations of mechanics of an elastic solid continuum (see Sec. 3).

The multi-particle-matrix system consists of isotropic cylindrical particles which are periodically distributed in an isotropic infinite matrix (see Fig. 1). The thermal stresses are a consequence of different thermal expansion coefficients of the isotropic matrix and isotropic particle (see Eq. (50)).

The infinite matrix is imaginarily divided into identical cubic cells (see Fig. 1), and the thermal stresses are thus investigated within the cubic cell (see Sec. 2). Each cell contains a central cylindrical particle. This multi-particle-matrix system represents a model system which is applicable to two-component materials of a precipitate-matrix type. The radius and the volume fraction of the cylindrical particles along with the cubic cell dimension represent microstructural parameters of the two-component material. The thermal stresses are then functions of these microstructural parameters. Numerical values for a real two-component material of the precipitate-matrix type (SiC-Si_3N_4 ceramics) are obtained.

Acknowledgments

This work was supported by the Slovak Grant Agency under the contract VEGA 2/0081/16, 2/0120/15, 2/0113/16 and by the Slovak Research and Development Agency under the contract APVV-0147-11.

References

[1] Greisel, M.; Jäger, J.; Moosburger-Will, J.; Sause, M.G.R.; Mueller, W.M.; Horn, S. *Composites: Part A* **2014,** *66,* 117-127.

[2] Meijer, G.; Ellyin, F.; Xia, Z. *Composites: Part B* **2000,** *31,* 29–37.

[3] Węglewski, W.; Basista, M.; Manescu, A.; Chmielewski, M.; Pietrzak, K.; and Schubert, Th. *Composites: Part B* **2014,** *67,* 119-124.

[4] Nabavi, S.M.; Ghajar, R. *Int. J. Eng. Sci.* **2010,** *48,* 1811-1823.

[5] Lee, J.; Hwang, K.-Y. *Int. J. Eng. Sci.* **1996,** *34,* 901–922.

[6] Tanigawa, Y.; Osako, M.K. *Int. J. Eng. Sci.* **1986,** *24,* 309–321.

[7] Abedian, A.; Szyszkowski, W.; Yannacopoulos, S. *Compos. Sci. Technol.* **1999,** *59,* 41–54.

[8] Ceniga, L. *Analytical Model of Thermal Stresses in Composite Materials I;* Nova Science Publishers: New York, US, 2008, pp 66–67, 136–137.

[9] Mizutani, T. *J. Mater. Sci.* **1996,** *11,* 483–494.

[10] Li, Sh.; Sauer, R.A.; Wang, G. *J. Appl. Mech.* **2007,** *74,* 770–783.

[11] Li, Sh.; Sauer, R.A.; Wang, G. *J. Appl. Mech.* **2007,** *74,* 784–797.

[12] Kushch, V.I. *Prikladnaja Mekhanika* **1985,** *21,* 18-27.

[13] Kushch, V.I. *Int. Appl. Mech.* **2004,** *40,* 893-899.

[14] Kushch, V.I. *Int. Appl. Mech.* **2004,** *40,* 1042-1049.

[15] Wu, Y.; Dong, Z. *Mater. Sci. Eng.* **1995,** *203,* 314–323.

[16] Sangani, A.S.; Mo, G. *J. Mech. Phys. Sol.* **1997,** *45,* 2001–2031.

[17] Bidulský, R.; Bidulská, J.; Grande, M.A. *Arch. Metallur. Mater.* **2013,** *58,* 365–370.

[18] Kvačkaj, T.; Kočisko, R.; Bidulský, R.; Bidulská, J.; Bella, P.; Lupták, M.; Kováčová, A.; Bacso, J. *Mater. Sci. For.* **2014,** *782,* 379–383.

[19] Bidulský, R.; Bidulská, J.; De Oro, R.; Hryha, E.; Maccarini, M.; Forno, I.; Grande, M.A. *Act. Phys. Polon. A* **2015,** *128,* 647–650.

[20] Mura, T. *Micromechanics of Defects in Solids;* Martinus Nijhoff Publishers: Dordrecht, NL, 1987, pp 1–3.

[21] Ceniga, L. *Analytical Models of Thermal Stresses in Composite Materials IV;* Nova Science Publishers: New York, US, 2015, pp 34–37.

[22] Brdička, M.; Samek, L.; Sopko, B. *Mechanics of Continuum;* Academia: Prague, CZ, 2000, pp 78–83.

[23] Trebuňa, F.; Šimčák, F.; Jurica, V. *Elasticity and Strength I;* Technical University: Košice, SK, 2005, pp 57–63.

[24] Trebuňa, F.;, Šimčák, F.; Jurica, V. *Examples and Problems of Elasticity and Strength I;* Technical University: Košice, SK, 2002, pp 134–138.

[25] Rektorys, K. *Review of Applied Mathematics;* SNTL: Prague, CZ, 1973, pp 253–267.

[26] Skočovský, P.; Bokůvka, O.; Palček, P. *Materials Science;* EDIS: Žilina, SK, 1996, pp 35–48.

In: Ceramic Materials
Editor: Jacqueline Perez
ISBN: 978-1-63485-965-3

Chapter 4

ANALYTICAL MODEL OF THERMAL STRESSES IN MULTI-COMPONENT POROUS MATERIALS

Ladislav Ceniga*
Institute of Materials Research, Slovak Academy of Sciences,
Košice, Slovak Republic

Abstract

This chapter deals with analytical modelling of thermal stresses which originate during a cooling process of an elastic solid continuum. This continuum consists of an isotropic infinite matrix with isotropic spherical particles and spherical pores. The particles and pores are both periodically distributed in the infinite matrix which is imaginarily divided into identical cubic cells. Each cell contains either a central particle or a central pore. This porous multi-particle-matrix system represents a model system which is applicable to porous two-component materials of a precipitate-matrix type characterized by microstructural parameters, i.e. the cubic cell dimension; radii and volume fractions of both the particles and of the pores. The thermal stresses are a consequence of different thermal expansion coefficients of the isotropic matrix and isotropic particle. Resulting from fundamental equations of mechanics of an elastic solid continuum, the thermal stresses are determined within this cell, and thus represent

*E-mail addresses: lceniga@yahoo.com, ladislavceniga65@gmail.com, lceniga@saske.sk, lceniga@imr.saske.sk

functions of these microstructural parameters. Finally, numerical values for a real porous two-component material of the precipitate-matrix type are obtained.

PACS: 46.25.Cc, 46.25.Hf, 46.70.Lk

Keywords: analytical modelling, thermal stress, composite material

1. Introduction

Thermal stresses which originate as a consequence of different thermal expansion coefficients of components of multi-component materials represent an important phenomenon in multi-component materials. These stresses are usually investigated by computational and experimental methods [1]–[7] are still of interest to materials scientists and engineers.

This chapter represents continuation of the author's book [8] which deals with the analytical modelling of the thermal stresses in two-component materials of a precipitate-matrix type with isotropic precipitates (isotropic particles) and an isotropic matrix. In contrast to [8], the analytical model of the thermal stresses in this chapter is determined for a porous isotropic matrix with spherical pores.

The thermal stresses originate during a cooling process at the temperature $T \in \langle T_f, T_r \rangle$, where T_f and T_r [9] are final and relaxation temperature of the cooling process, respectively. As defined in [9], the relaxation temperature T_r is such temperature below that the stress relaxation as a consequence of thermal-activated processes does not occur in a material. The relaxation temperature is defined approximately by the relationship $T_r = (0.35 - 0.4) \times T_m$ [9] and exactly by an experiment, where T_m is a melting point of a two-component material.

If precipitates are formed from a liquid matrix of the two-component material, then the melting point T_m represents minimum of the set $\{T_{mp}, T_{mm}\}$, where T_{mp} and T_{mm} are melting points of the precipitate (particle) and the matrix [9], respectively. If the precipitates are formed from a solid matrix, then T_m represents a melting point of the two-component material [9].

The thermal stresses which originate at the temperature $T \in \langle T_f, T_r \rangle$ are thus a consequence of the condition $\beta_m - \beta_p \neq 0$. The coefficient β_q is derived as

$$\beta_q = \int_T^{T_r} \alpha_q \, dT, \quad q = p, m, \tag{1}$$

where α_q is a thermal expansion coefficient of the isotropic spherical particle ($q = p$) and the isotropic matrix ($q = m$).

Within the analytical modelling presented in this chapter, this cooling process is characterized by a homogeneous temperature change. Using the spherical coordinates (r, φ, ν), the homogeneous temperature change is then characterized by the condition $\partial T/\partial r = \partial T/\partial \varphi = \partial T/\partial \nu = 0$.

With regard to materials engineering, experimental results are usually compared with the Selsing's analytical model [10]–[23] which determines thermal stresses in a one-particle-matrix system with an isotropic spherical particle embedded in an isotropic infinite matrix. On the one hand, the Selsing's analytical model does not consider neither the inter-particle distance d nor the particle volume fraction $v_p = V_p/V_m$, where V_p and V_m is volume of the particle and the matrix, respectively, and then $d \to \infty$, $V_m \to \infty$, $v_p = 0$ for the Selsing's analytical model. On the other hand, in spite of this fact, the Selsing's analytical model which was determined in 1961 is still used by materials scientists at present [16]–[23].

As a continuation of the Selsing's analytical model, Mizutani [24] determined an analytical model for the thermal stresses for a multi-particle-matrix system with isotropic components. The Mizutani's analytical model considers an imaginary division of the isotropic infinite matrix into identical spherical cells with a central isotropic spherical particle with the radius R_p. The thermal stresses which are investigated within this cell are then functions of the microstructural parameters d, v_p, R_p. On the one hand, the influence of the thermal stresses of all cells on the thermal stresses in a certain cell as well as the influence of the thermal stresses of neighbouring cells on the thermal stresses in a certain cell can be hardly expressed by an exact mathematical solution. On the other hand, the analytical model of the thermal stresses which is determined within a cell is generally considered to represent more than sufficient mathematical solution for materials scientists and engineers. Using final mathematical formulae for the thermal stresses, materials scientists and engineers can then determine numerical values of the thermal stresses for a real two-component material.

As presented in [24]–[31], this 'cell approach' is used within mathematical procedures applied to analytical modelling of periodic model systems with an infinite matrix. The infinite matrix of the periodic model systems can be then imaginarily divided into e.g. spherical cells [24]–[26], cubic cells [27, 28], however, in contrast to cubic cells, two spherical cells are mutually connected in one point [24]–[26]. The matrix between two spherical cells is not considered, and accordingly, cubic cells which are considered in this chapter are preferable.

Finally, as presented in [24, 32], to obtain analytical solutions, real two-component materials with finite dimensions and with aperiodically distributed particles (precipitates) are then replaced by a model system which consists of an infinite matrix with periodically distributed particles of a defined shape (e.g. spherical particles). Additionally, the case when an infinite matrix is considered within the analytical modelling is of particular interest for the mathematical simplicity of analytical solutions. As analysed in [32], such analytical solutions are assumed to exhibit sufficient accuracy due to the size of material components (e.g. precipitates) which is relatively small in comparison with the size of macroscopic material samples, macroscopic structural elements, etc.

2. Cell Model

With regard to the analytical modelling, real porous two-component materials with isotropic components of finite dimensions are replaced by an infinite porous multi-particle-matrix system (see Fig. 1). The porous multi-particle-matrix system represents a model system. The model system is represented by isotropic spherical particles which are periodically distributed in a porous isotropic infinite matrix with periodically distributed spherical pores. The porous infinite matrix is imaginarily divided into identical cubic cells with a central spherical particle and into identical cubic cells with a central spherical pore.

The cubic cell represents such part of the porous multi-particle-matrix system which is related to one spherical particle (see the cubic cell *1234* in Fig. 1) and to one spherical pore (see the cubic cell *2567* in Fig. 1). Additionally, as presented in Fig. 1, one cubic cell with the central spherical pore (the cube *2567*) is related to 26 cubic cells with the central spherical particle (see the cube *48910* in Fig. 1).

The inter-particle distance d, the particle radius R_p, the pore radius R_v (pore=void), the particle volume fraction v_p and the pore volume fraction v_v, which represent parameters of the cubic cells with the spherical particle and with the spherical pore, are microstructural parameters of porous two-component materials of a precipitate-matrix type, where v_p and v_v have the forms

$$v_p = \frac{26V_p}{27V_c} = \frac{104\pi}{81}\left(\frac{R_p}{d}\right)^3 \in \left(0, v_{pmax}\right\rangle, \quad v_{pmax} = \frac{13\pi}{81}, \tag{2}$$

$$v_v = \frac{V_v}{27V_c} = \frac{4\pi}{81}\left(\frac{R_v}{d}\right)^3 \in \left(0, v_{fmax}\right\rangle, \quad v_{pmax} = \frac{\pi}{162}, \tag{3}$$

$$\frac{v_p}{v_v} = 26\left(\frac{R_p}{R_v}\right)^3, \tag{4}$$

where $V_p = 4\pi R_p^3/3$, $V_v = 4\pi R_v^3/3$ and $V_c = d^3$ is volume of the spherical particle, the spherical pore and the cubic matrix, respectively. The maximum values v_{pmax} and v_{fmax} result from the condition $R_p = d/2$ and $R_v = d/2$, respectively.

In case of a real porous two-component material of the precipitate-matrix type, these microstructural parameters are obtained by experimental techniques. The thermal stresses are investigated within the cubic cell with the spherical particle and within the cubic cell with the pore, and thus represent functions of these microstructural parameters. Additionally, the surface of the cubic cell defines positions for which one of boundary conditions for the infinite matrix is determined (see Eq. (57)).

The thermal stresses, as functions of the spherical coordinates $[r, \varphi, \nu]$, are derived in the Cartesian system $(Px_1'x_2'x_3')$ (see Fig. 2). The spherical coordinates (r, φ, ν) are used due to the spherical shapes of the particles and of the pores of the porous multi-particle-matrix system. The axes x_1' and x_2', x_3' represent radial and tangential directions, respectively, and $r = \left|\overline{OP}\right|$, where O is a centre of the spherical particle. Due to the symmetry of the cubic cell as well as due to the isotropy of the porous multi-particle-matrix system, the thermal stresses are sufficient to be investigated within one eighth of the cubic cell, i.e. for $\varphi, \nu \in \langle 0, \pi/2\rangle$, where $r \in \langle 0, R_p\rangle$ for the spherical particle and $r \in \langle R_p, r_c\rangle$ for the cell matrix, respectively.

The distance $r_c = r_c(\varphi, \nu) = \left|\overline{OC_3}\right|$ for $\nu \in \langle 0, \nu^*)$ and $r_c = r_c(\varphi, \nu) = \left|\overline{OC_1}\right|$ for $\nu \in \langle \nu^*, \pi/2\rangle$ (see Fig. 3) has the form [8]

$$r_c = d\, f_c = \frac{R_p\, f_c}{3}\left(\frac{104\pi}{3v_p}\right)^{1/3} = \frac{R_v\, f_c}{3}\left(\frac{4\pi}{3v_v}\right)^{1/3}. \tag{5}$$

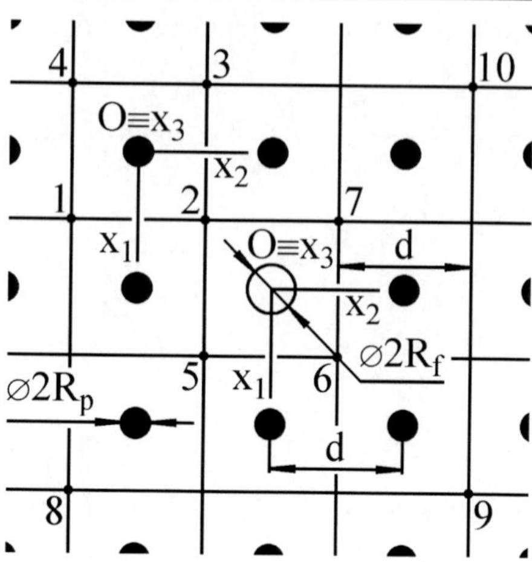

Figure 1. The porous multi-particle-matrix system imaginarily divided into identical cubic cells with a central spherical particle (see the cubic cell *1234*) and with a central spherical pore (see the cubic cell *2567*) both in the point O of the Cartesian system $(Ox_1x_2x_3)$, where a dimension of the cubic cell is identical to the inter-particle distance d; R_p and R_v are radii of the particle and the pore, respectively. One cubic cell with the central spherical pore is related to 26 cubic cells with the central spherical particle (see the cube *2567*).

The function $f_c = f_c(\varphi, \nu)$ along with the coefficient c_φ and the angle ν^* (see Fig. 3) are derived as

$$f_c = \frac{\sqrt{1 + c_\varphi^2}}{2\left[c_\varphi \cos(\nu^* - \nu) + \sin(\nu^* - \nu)\right]}, \quad \nu \in (0, \nu^*\rangle, \tag{6}$$

$$f_c = \frac{1}{2c_\varphi \sin\nu}, \quad \nu \in \left\langle \nu^*, \frac{\pi}{2} \right\rangle, \tag{7}$$

$$c_\varphi = \cos\varphi, \quad \varphi \in \left\langle 0, \frac{\pi}{4} \right\rangle, \tag{8}$$

$$c_\varphi = \sin\varphi, \quad \varphi \in \left(\frac{\pi}{4}, \frac{\pi}{2} \right\rangle, \tag{9}$$

$$\nu^* = \arctan\left(\frac{1}{c_\varphi} \right). \tag{10}$$

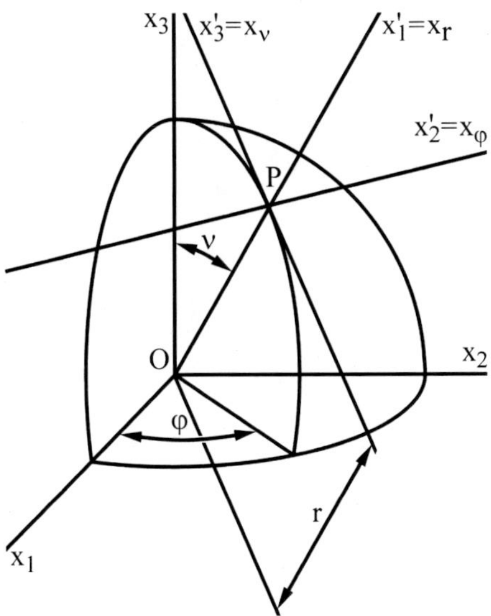

Figure 2. The axes $x_1' = x_r = OP$ and $x_2' = x_\varphi \parallel x_1 x_2$, $x_3' = x_\nu$ defining radial and tangential directions regarding the Cartesian system $(Px_1' x_2' x_3')$, respectively, and the arbitrary point P with a position determined by the spherical coordinates $[r, \varphi, \nu]$ regarding the Cartesian system $(Ox_1 x_2 x_3)$ (see Fig. 1). O is a centre of the spherical particle, and x_2', x_3' are tangents to a surface of a sphere with the radius $r = \left|\overline{OP}\right|$ representing length of the abscissa $|OP|$.

The boundary condition for a surface of the cell with the spherical particle (see the cell *1234*) given by Eq. (57), along with the formulae for the thermal stresses given by (24)–(27), (63) and with the boundary conditions given by Eqs. (54), Eq. (56) result in a dependence of the thermal stresses in the spherical particle and the cell matrix on the microstructural parameters R_p, d or R_p, v_p (see Eqs. (2)).

3. Fundamental Equations

The analytical modelling presented in this chapter results from fundamental equations of solid continuum mechanics. These fundamental equations are represented by the Cauchy's equations, the equilibrium equations and the Hooke's

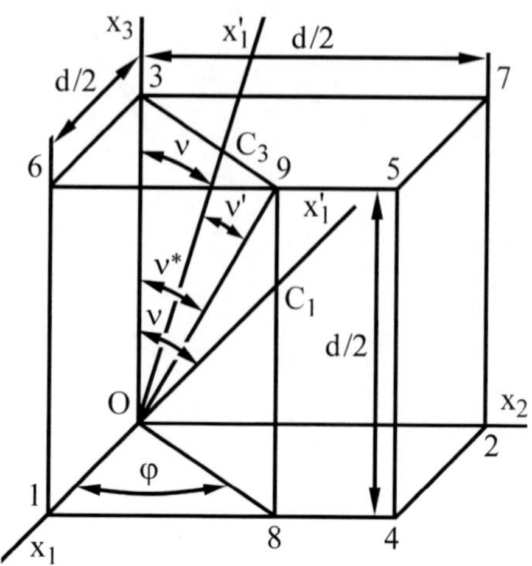

Figure 3. The points C_3 ($r_c = |\overline{OC_3}|$) and C_1 ($r_c = |\overline{OC_1}|$) as points of intersection of the axis x_1', representing a radial direction, with the surfaces *3657* and *1456* of one eighth of the cubic cell (see Fig. 1) for $\nu \in \langle 0, \nu^*)$ and $\nu \in \langle \nu^*, \pi/2 \rangle$, respectively.

law for an isotropic elastic solid continuum. The Cauchy's equations and the equilibrium equations for the infinitesimal spherical cap in the arbitrary point P with a position determined by the spherical coordinates $[r, \varphi, \nu]$ (see Fig. 4) regarding the Cartesian system $(Ox_1x_2x_3)$ are determined for the Cartesian system $(Px_1'x_2'x_3')$ (see Fig. 2). The Cauchy's equations which represent geometric equations define relationships between strains and displacements of an infinitesimal part of a solid continuum. The equilibrium equations which are related to the axes x_1', x_2', x_3' (see Fig. 2) are based on a condition of the equilibrium of forces which act on sides of this infinitesimal part.

The infinitesimal spherical cap in the arbitrary point P (see Fig. 4) is described by the dimensions dr, $rd\varphi$,$rd\nu$ along the axes x_1', x_2', x_3' of the Cartesian system $(Px_1'x_2'x_3')$, respectively. The axis x_1' represents a normal of the surfaces S_r and S_{r+dr} with the surface area $A_r = r^2\, d\varphi\, d\nu$ and $A_{r+dr} = (r+dr)^2\, d\varphi\, d\nu$ at the radii r and $r + dr$, respectively. As analysed in [33], this infinitesimal spherical cap exhibits the radial displacement u_1' along the axis x_1'. Accordingly, the Cauchy's equations for the radial strain ε_{11}' along the axis x_1', tangential strain ε_{ii}' and the shear strains $\varepsilon_{1i}' = \varepsilon_{i1}'$ along the axis x_i'

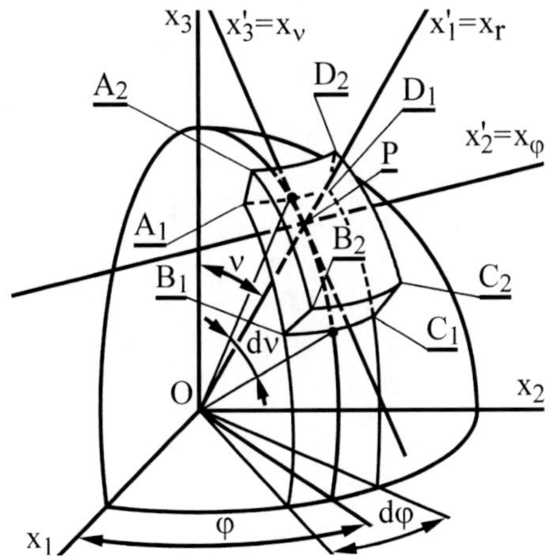

Figure 4. The infinitesimal spherical cap at the point P (see Fig. 2) with the surfaces $S_r = A_1B_1C_1D_1$ and $S_{r+dr} = A_2B_2C_2D_2$ at the radii $r = \left|\overline{OP}\right|$ and $r + dr$, respectively. Dimensions of the infinitesimal spherical cap are as follows: $|A_1A_2| = |B_1B_2| = |C_1C_2| = |D_1D_2| = dr$, $|A_1D_1| = |B_1C_1| = r \times d\varphi$, $|A_1B_1| = |C_1D_1| = r \times d\nu$, $|A_2D_2| = |B_2C_2| = (r + dr) \times d\varphi$, $|A_2B_2| = |C_2D_2| = (r + dr) \times d\nu$.

(i = 2,3) are derived as [33]

$$\varepsilon'_{11} = \frac{\partial u'_1}{\partial r}, \tag{11}$$

$$\varepsilon'_{22} = \varepsilon'_{33} = \frac{u'_1}{r}, \tag{12}$$

$$\varepsilon'_{12} = \varepsilon'_{21} = \frac{1}{r}\frac{\partial u'_1}{\partial \varphi}, \tag{13}$$

$$\varepsilon'_{13} = \varepsilon'_{31} = \frac{1}{r}\frac{\partial u'_1}{\partial \nu}, \tag{14}$$

Consequently, the equilibrium equations for the radial stress σ'_{11} acting along the axis x'_1, tangential stresses σ'_{ii} and the shear stresses $\sigma'_{1i} = \sigma'_{i1}$ acting along the axis x'_i (i = 2,3) have the forms [33]

$$2\sigma'_{11} - \sigma'_{22} - \sigma'_{33} + r\frac{\partial \sigma'_{11}}{\partial r} + \frac{\partial \sigma'_{12}}{\partial \varphi} + \frac{\partial \sigma'_{13}}{\partial \nu} = 0, \tag{15}$$

$$\frac{\partial \sigma'_{22}}{\partial \varphi} + 3\sigma'_{12} + r\,\frac{\partial \sigma'_{12}}{\partial r} = 0, \tag{16}$$

$$\frac{\partial \sigma'_{33}}{\partial \nu} + 3\sigma'_{13} + r\,\frac{\partial \sigma'_{13}}{\partial r} = 0. \tag{17}$$

With regard to the shear strain $\varepsilon'_{23} = \varepsilon'_{32} \propto [(\partial u'_2/\partial \nu) + (\partial u'_3/\partial \varphi)]$ in the Cartesian system $(Px'_1x'_2x'_3)$ (see Fig. 2) [34], we get $\varepsilon'_{23} = 0$ due to $u'_2 = 0$, $u'_3 = 0$, where u'_2 and u'_3 represent displacements of the infinitesimal spherical cap along the axes x'_2 and x'_3 (see Fig. 2), respectively, i.e. along tangential directions. Similarly, with regard to the Hooke's law for an isotropic elastic solid continuum, i.e. $\varepsilon'_{23} = s_{44}\sigma'_{23}$ (see Eq. (23)). [34], we get $\sigma'_{23} = 0$.

Finally, with regard to $\varepsilon'_{23} = 0$, $\sigma'_{23} = 0$, the Hooke's law for an isotropic elastic solid continuum is derived as [34]–[36]

$$\varepsilon'_{11} = s_{11}\sigma'_{11} + s_{12}\left(\sigma'_{22} + \sigma'_{33}\right), \tag{18}$$

$$\varepsilon'_{22} = s_{12}\left(\sigma'_{11} + \sigma'_{33}\right) + s_{11}\sigma'_{22}, \tag{19}$$

$$\varepsilon'_{33} = s_{12}\left(\sigma'_{11} + \sigma'_{22}\right) + s_{11}\sigma'_{33}, \tag{20}$$

$$\varepsilon'_{13} = s_{44}\sigma'_{13}, \tag{21}$$

$$\varepsilon'_{12} = s_{44}\sigma'_{12}. \tag{22}$$

The elastic moduli s_{11}, s_{12}, s_{44} which represent functions of the Young's modulus E and Poisson's ratio μ have the forms [34]–[36]

$$s_{11} = \frac{1}{E}, \quad s_{12} = -\,\frac{\mu}{E}, \quad s_{44} = \frac{2\,(1+\mu)}{E}, \tag{23}$$

The Young' modulus E and the Poisson's ratio μ are related to the spherical particle ($q = p$) and the cell matrix ($q = m$). Consequently, the transformations $E \to E_q$, $\mu \to \mu_q$ ($i, j = 1, \ldots, 6$; $q = p,m$) are required to be considered.

4. Mathematical Procedures

Let the Cauchy's equations (see Eqs. (11)–(14)) be substituted to the Hooke's law for an isotropic solid elastic continuum (see Eqs. (18)–(23)). Consequently, the radial stress σ'_{11} acting along the axis x'_1 (see Fig. 2), the tangential stress σ'_{22} acting along the axis x'_2, the tangential stress σ'_{33} acting along the axis x'_3, the shear stress σ'_{12} acting along the axis x'_2 and the shear stress σ'_{13} acting along the axis x'_3 have the forms

$$\sigma'_{11} = (c_1 + c_2)\frac{\partial u'_1}{\partial r} - 2c_2\frac{u'_1}{r}, \tag{24}$$

$$\sigma'_{22} = \sigma'_{33} = -c_2\frac{\partial u'_1}{\partial r} + c_1\frac{u'_1}{r}, \tag{25}$$

$$\sigma'_{12} = \frac{1}{s_{44}r}\frac{\partial u'_1}{\partial \varphi}, \tag{26}$$

$$\sigma'_{13} = \frac{1}{s_{44}r}\frac{\partial u'_1}{\partial \nu}, \tag{27}$$

where $\sigma'_{ij} = \sigma'_{ji}$ $(i, j = 1,2,3)$ [34]–[36], and the condition $\sigma'_{22} = \sigma'_{33}$ results from the condition $\varepsilon'_{22} = \varepsilon'_{33}$ (see Eqs. (12), (19), (20)). The coefficients c_1, c_2, c_3 are derived as (see Eq. (23))

$$c_1 = \frac{s_{11}}{s_{11}(s_{11} + s_{12}) - 2s_{12}^2} = \frac{E}{(1+\mu)(1-2\mu)},$$
$$c_2 = \frac{s_{12}\,c_1}{s_{11}} = -\frac{\mu E}{(1+\mu)(1-2\mu)}, \quad c_3 = \frac{s_{44}c_1 + 2}{s_{44}c_2 - 1} = -4(1-\mu). \tag{28}$$

In case of an isotropic elastic solid continuum, we get $\mu = 0.25$ [34]. In case of a real isotropic material, we get $\mu < 0.5$ [9], and then $c_3 < 0$. The coefficient c_3 is considered within the following mathematical procedures.

Additionally, with regard to Eqs. (11)–(14), (24)–(27), the elastic energy density w [34]–[36] has the form

$$w = \sum_{i,j=1}^{3} \varepsilon'_{ij}\sigma'_{ij} = \frac{1}{2}(c_1 + c_2)\left(\frac{\partial u'_1}{\partial r}\right)^2 - \frac{2c_1\,u'_1}{r}\frac{\partial u'_1}{\partial r} + \frac{c_1\,(u'_1)^2}{r^2}$$
$$+ \frac{1}{s_{44}\,r^2}\left[\left(\frac{\partial u'_1}{\partial \varphi}\right)^2 + \left(\frac{\partial u'_1}{\partial \nu}\right)^2\right], \tag{29}$$

where $\varepsilon'_{23} = 0$, $\sigma'_{23} = 0$ (see Sec. 3).

Let the equations (24)–(27) be substituted to Eq. (15) as well as to the sum $[\partial\text{Eq.}(16)/\partial\varphi]$ + $[\partial\text{Eq.}(18)/\partial\nu]$. The equilibrium equations (15)–(17) are thus transformed to the forms

$$r^2\frac{\partial^2 u'_1}{\partial r^2} + 2r\frac{\partial u'_1}{\partial r} - 2u'_1 + \frac{U_1}{s_{44}(c_1 + c_2)} = 0, \tag{30}$$

$$r\,\frac{\partial U_1}{\partial r} = c_3\,U_1, \tag{31}$$

where the function U_1 is derived as

$$U_1 = \frac{\partial^2 u_1'}{\partial \varphi^2} + \frac{\partial^2 u_1'}{\partial \nu^2}. \tag{32}$$

A solution of this system of the differential equations (30), (31) is determined by the following mathematical procedures which result in a solution for the radial displacement u_1' (see Eqs. (38)–(40)) along the axis x_1' (see Fig. 2).

Performing $r\,[\partial \text{Eq.}(31)/\partial r]$, we get

$$r^2\,\frac{\partial^2 U_1}{\partial r^2} + (1 - c_3)\,r\,\frac{\partial U_1}{\partial r} = 0. \tag{33}$$

Substituting Eq. (31) to Eq. (33), we get

$$r^2\,\frac{\partial^2 U_1}{\partial r^2} + c_3\,(1 - c_3)\,U_1 = 0. \tag{34}$$

If the dependence $U_1 - r$ is assumed in the form $U_1 = r^\lambda$, then the solution U_1 of the differential equation (34) has the form

$$U_1 = \frac{\partial^2 u_1'}{\partial \varphi^2} + \frac{\partial^2 u_1'}{\partial \nu^2} = \sum_{i=1}^{2} C_i\,r^{\lambda_i}, \tag{35}$$

where C_1, C_2 are integration constants which are related to the mathematical procedure 1. The exponent λ_i $(i = 1{,}2)$ is derived as

$$\lambda_i = \frac{1}{2}\left[1 + (\delta_{1i} - \delta_{2i})\,\sqrt{D}\right], \quad i = 1, 2,$$
$$D = 1 - 4\,c_3\,(1 - c_3) = 1 + 16\,(1 - \mu)\,[1 + 4\,(1 - \mu)], \tag{36}$$

Due to $\mu < 0.5$ for a real isotropic material [9], and $c_3 < 0$ (see Eq. (28)), we get $D > 0$, and the real exponents $\lambda_1 > 3$, $\lambda_2 < -2$.

Considering Eq. (35), the differential equation (30) is transformed to the form

$$\frac{\partial^2 u_1'}{\partial r^2} + \frac{2}{r}\,\frac{\partial u_1'}{\partial r} - \frac{2u_1'}{r^2} = -\,\frac{1}{s_{44}\,(c_1 + c_2)}\sum_{i=1}^{2} C_i\,r^{\lambda_i - 2}. \tag{37}$$

Using the Wronskian's method [37], the radial displacement $u'_{1q} = u'_{1q}(r, \varphi, \nu)$ in the spherical particle ($q = p$) and the cell matrix ($q = m$) has the form

$$u'_{1q} = \sum_{i=1}^{2} C_{iq}\, \xi_{iq}\, u_{irq}, \quad q = p, m. \tag{38}$$

The function $u_{irq} = u_{irq}(r)$ and the coefficient ξ_{iq} ($i = 1,2$; $q = p,m$) have the forms

$$u_{irq} = r^{\lambda_{iq}}, \quad i = 1, 2; \quad \lambda_{1q} > 3, \quad \lambda_{2q} < -2; \quad q = p, m, \tag{39}$$

$$\xi_{iq} = \frac{1}{3 s_{44}\,(c_1 + c_2)} \left(\frac{1}{\lambda_{iq} + 2} - \frac{1}{\lambda_{iq} - 1} \right), \quad i = 1, 2; \quad q = p, m, \tag{40}$$

where $u_{1rq} = u_{1rq}(r)$ and $u_{2rq} = u_{2rq}(r)$ are increasing and decreasing functions of the variable r due to $\lambda_{1q} > 3$ and $\lambda_{2q} < -2$, respectively.

With regard to Eqs. (11)–(14), (24)–(27), we get

$$\varepsilon'_{11q} = \sum_{i=1}^{2} C_{iq}\, \xi_{iq}\, \lambda_{iq}\, r^{\lambda_{iq}-1}, \tag{41}$$

$$\varepsilon'_{22q} = \varepsilon'_{33q} = \sum_{i=1}^{2} C_{iq}\, \xi_{iq}\, r^{\lambda_{iq}-1}, \tag{42}$$

$$\varepsilon'_{12q} = \sum_{i=1}^{2} \frac{\partial C_{iq}}{\partial \varphi}\, \xi_{iq}\, r^{\lambda_{iq}-1}, \tag{43}$$

$$\varepsilon'_{13q} = \sum_{i=1}^{2} \frac{\partial C_{iq}}{\partial \nu}\, \xi_{iq}\, r^{\lambda_{iq}-1}, \tag{44}$$

$$\sigma'_{11q} = \sum_{i=1}^{2} C_{iq}\, \xi_{iq}\, \xi_{2+iq}\, r^{\lambda_{iq}-1}, \tag{45}$$

$$\sigma'_{22q} = \sigma'_{33q} = \sum_{i=1}^{2} C_{iq}\, \xi_{iq}\, \xi_{4+iq}\, r^{\lambda_{iq}-1}, \tag{46}$$

$$\sigma'_{12q} = \frac{1}{s_{44q}} \sum_{i=1}^{2} \frac{\partial C_{iq}}{\partial \varphi}\, \xi_{iq}\, r^{\lambda_{iq}-1}, \tag{47}$$

$$\sigma'_{13q} = \frac{1}{s_{44q}} \sum_{i=1}^{2} \frac{\partial C_{iq}}{\partial \nu} \, \xi_{iq} \, r^{\lambda_{iq}-1}, \tag{48}$$

where the coefficients ξ_{2+iq}, ξ_{4+iq} ($i = 1,2$) are derived as

$$\xi_{2+iq} = \lambda_{iq} \left(c_{1q} + c_{2q} \right) - 2c_{2q}, \quad i = 1, 2, \tag{49}$$

$$\xi_{4+iq} = c_{1q} - \lambda_{iq} \, c_{2q}, \quad i = 1, 2. \tag{50}$$

With regard to Eqs. (11)–(14), (24)–(27), (41)–(50), the thermal-stress induced elastic energy density w_q [34]–[36] has the form

$$\begin{aligned} w_q &= \sum_{i,j=1}^{3} \varepsilon'_{ijq} \sigma'_{ijq} = \frac{1}{2} \left(c_{1q} + c_{2q} \right) \left(\frac{\partial u'_{1q}}{\partial r} \right)^2 - \frac{2 c_{1q} \, u'_{1q}}{r} \frac{\partial u'_{1q}}{\partial r} + \frac{c_{1q} \left(u'_{1q} \right)^2}{r^2} \\ &+ \frac{1}{s_{44q} \, r^2} \left[\left(\frac{\partial u'_{1q}}{\partial \varphi} \right)^2 + \left(\frac{\partial u'_{1q}}{\partial \nu} \right)^2 \right] \\ &= \sum_{i,j=1}^{2} \xi_{2+iq} \, \xi_{jq} \left[C_{iq} \, C_{jq} \, \xi_{4+i+2jq} + \frac{1}{s_{44q}} \left(\frac{\partial C_{iq}}{\partial \varphi} \frac{\partial C_{jq}}{\partial \varphi} + \frac{\partial C_{iq}}{\partial \nu} \frac{\partial C_{jq}}{\partial \nu} \right) \right] r^{\lambda_{iq} - \lambda_{jq} - 2}, \end{aligned} \tag{51}$$

where the coefficient $\xi_{4+i+2jq}$ ($i, j = 1,2$) is derived as

$$\xi_{4+i+2jq} = \lambda_{iq} \left[\frac{\lambda_{jq}}{2} \left(c_{1q} + c_{2q} \right) - 2c_{2q} \right], \quad i, j = 1, 2. \tag{52}$$

The thermal-stress induced elastic energy W_q which is accumulated in the spherical particle ($q = p$), in the cell matrix around the particle ($q = m$) and in the cell matrix around the pore (=flaw, $q = v$) has the form

$$W_q = 8 \int_0^{\pi/2} \int_0^{\pi/2} \int_{r_1}^{r_2} w_q \, r^2 \, dr \, d\varphi \, d\nu = 8 \int_0^{\pi/2} \int_0^{\pi/2} \int_{r_1}^{r_2} \left(\frac{1}{2} \sum_{i,j=1}^{3} \varepsilon'_{ijq} \sigma'_{ijq} \right) r^2 \, dr \, d\varphi \, d\nu. \tag{53}$$

The boundaries $r_1 = 0$, $r_2 = R_p$; $r_1 = R_p$, $r_2 = r_c$ (see Eqs. (5)–(10)); and $r_1 = R_v$, $r_2 = r_c$ are valid for the spherical particle; for the cell matrix around the spherical particle; and for the cell matrix around the pore, respectively.

5. Boundary Conditions

5.1. Spherical Particle

The absolute value $\left|u'_{1p}\right|$ is required to represent an increasing function of $r \in \langle 0, R_p \rangle$, exhibiting a maximum value on the particle-matrix boundary (i.e. for $r = R_p$). Additionally, due to $\sigma'_{iip} \propto r^{\lambda_p - 1}$, $\sigma'_{12+jp} \propto r^{\lambda_p - 1}$ ($i = 1,2,3$; $j = 0,1$; see Eqs. (24)–(27)), the exponent λ_p is required to be $\lambda_p > 1$, otherwise $(\sigma_{iip})_{r \to 0} \to \pm\infty$, $(\sigma_{12+jp})_{r \to 0} \to \pm\infty$.

With regard to Eqs. (38), (39), $u_{2rp} = r^{\lambda_{2p}}$ is a decreasing function of the variable r due to $\lambda_{2p} < -2$, and then we get $C_{2p} = 0$.

With regard to $\lambda_{1p} > 3$ (see Eq. (39)), the boundary condition for the determination of the integration constant C_{1p} (see Eq. (38)) has the form [8]

$$\left(\sigma'_{11p}\right)_{r=R_p} = -p, \tag{54}$$

where the determination of the radial stress p acting at the particle-matrix boundary is analysed in Sec. 5.4.

With regard to Eqs. (45), (54), we get

$$C_{1p} = -\frac{p}{\xi_{1p}\, \xi_{3p}\, R_p^{\lambda_{1p}-1}} \tag{55}$$

5.2. Cell Matrix Around Particle

The absolute value $|u'_{1m}|$ which exhibits a maximum value on the particle-matrix boundary (i.e. for $r = R_p$) is required to represent a decreasing function of $r \in \langle R_p, r_c \rangle$, where r_c is given by Eqs. (5)–(10). The decreasing course of the dependence $|u'_{1m}| - r$ is ensured by the integration constants C_{1m}, C_{2m} in Eq. (38). The boundary conditions for the determination of the integration constants C_{1m}, C_{2m} are derived as

$$\left(\sigma'_{11m}\right)_{r=R_p} = -p, \tag{56}$$

$$\left(u'_{1m}\right)_{r=r_c} = u'_{1c} = u_{1c0}\, f_c. \tag{57}$$

where the function $f_c = f_c\,(\varphi, \nu)$ is given by Eqs. (6)–(10), and u_{1c0} is a radial displacement along the axis x_i ($i = 1,2,3$) on the cubic cell surface (i.e. for $r = d/2$).

The determination of the radial stress p acting at the particle-matrix boundary and the determination of the radial displacement u_{1c0} are analysed in Secs. 5.4 and 5.5, respectively.

With regard to Eqs. (38)–(40), (45), (56), (57), we get

$$\begin{aligned} C_{im} &= -\frac{1}{\xi_{im}\, r_c^{\lambda_{im}-1}}\left(\frac{p}{\zeta_{im}} + \frac{u'_{1c}\,\xi_{5-im}}{r_c\,\zeta_{2+im}}\right) \\ &= -\frac{1}{\xi_{im}\, r_c^{\lambda_{im}-1}}\left(\frac{p}{\zeta_{im}} + \frac{u_{1c0}\,\xi_{5-im}}{d\,\zeta_{2+im}}\right), \quad i = 1, 2, \end{aligned} \tag{58}$$

where $u'_{1c}/r_c = u_{1c0}/d$ (see Eqs. (2), (57)).

The coefficients ζ_{im}, ξ_{4+2im} ($i = 1,2$) have the forms

$$\zeta_{im} = \xi_{2+im}\left(\frac{R_p}{r_c}\right)^{\lambda_{im}-1} - \xi_{5-im}\left(\frac{R_p}{r_c}\right)^{\lambda_{3-im}-1}, \quad i = 1, 2, \tag{59}$$

$$\zeta_{2+im} = \xi_{2+im}\left(\frac{R_p}{r_c}\right)^{\lambda_{im}-\lambda_{3-im}} - \xi_{5-im}, \quad i = 1, 2. \tag{60}$$

5.3. Cell Matrix Around Pore

The absolute value $|u'_{1v}|$ is required to represent an increasing function of $r \in \langle R_v, r_c \rangle$, exhibiting a maximum value on the cell surface (i.e. for $r = r_c$).

With regard to Eqs. (38), (39), $u_{2rv} = r^{\lambda_{2v}}$ is a decreasing function of the variable r due to $\lambda_{2v} < -2$, and then we get $C_{2v} = 0$.

With regard to $\lambda_{1v} > 3$ (see Eq. (39)), the boundary condition for the determination of the integration constant C_{1v} (see Eq. (38)) has the form

$$\left(u'_{1v}\right)_{r=r_c} = -u'_{1c} = -u_{1c0}\, f_c. \tag{61}$$

where the determination of the radial displacement u_{1c0} is analysed in Sec. 5.5.

With regard to Eqs. (38), (39), (61), we get

$$C_{1v} = -\frac{u'_{1c}}{\xi_{2m}\, r_c^{\lambda_{2m}}} = -\frac{u_{1c0}}{d\,\xi_{2m}\, r_c^{\lambda_{2m}-1}}, \tag{62}$$

where $u'_{1c}/r_c = u_{1c0}/d$ (see Eqs. (2), (57)).

5.4. Determination of Radial Stress p

The thermal stresses in the porous multi-particle-matrix system (see Fig. 1) are a consequence of the radial stress p which acts at a particle-matrix boundary. The compressive or tensile radial stress, $p > 0$ or $p < 0$, results from the difference $\beta_m - \beta_p > 0$ or $\beta_m - \beta_p < 0$ (see Eq. (1)). As analysed in [8], the radial stress p is determined by the following formula

$$\left(\varepsilon'_{22m}\right)_{r=R_p} - \left(\varepsilon'_{22p}\right)_{r=R_p} = \beta_m - \beta_p. \tag{63}$$

5.5. Determination of u_{1c0}

The radial displacement u_{1c0} along the axis x_i ($i = 1,2,3$) on the cubic cell surface (i.e. for $r = d/2$) is determined by the following energy analysis.

The thermal-stress induced elastic energy of the multi-particle-matrix system with pores (see Fig. 1) is represented by both the elastic energy W_v and by the elastic energy $26\left(W_p + W_m\right)$ accumulated in 26 cubic cells with particles. These 26 cubic cells are related to one porous cubic cell. Stresses acting in the porous cubic cell are induced by the thermal stresses acting in these 26 cubic cells. The stresses acting in both the 26 cubic cells and the porous cubic cell are required to fulfilled the following energy condition

$$\frac{\partial\left[26\left(W_p + W_m\right) - W_v\right]}{\partial u_{1c0}} = 0 \tag{64}$$

The condition (64) is considered for the determination of the radial displacement u_{1c0}.

6. Thermal Stresses in Porous Multi-Particle-Matrix System

6.1. Spherical Particle

With regard to Eqs. (41)–(48), (55), we get

$$u'_{1p} = -\frac{pr}{\xi_{3p}}\left(\frac{r}{R_p}\right)^{\lambda_{1p}-1}, \tag{65}$$

$$\varepsilon'_{11p} = -\frac{p\lambda_{1p}}{\xi_{3p}}\left(\frac{r}{R_p}\right)^{\lambda_{1p}-1}, \tag{66}$$

$$\varepsilon'_{22p} = \varepsilon'_{33p} = -\frac{p}{\xi_{3p}} \left(\frac{r}{R_p}\right)^{\lambda_{1p}-1}, \tag{67}$$

$$\varepsilon'_{12p} = -\frac{1}{\xi_{3p}} \frac{\partial p}{\partial \varphi} \left(\frac{r}{R_p}\right)^{\lambda_{1p}-1}, \tag{68}$$

$$\varepsilon'_{13p} = -\frac{1}{\xi_{3p}} \frac{\partial p}{\partial \nu} \left(\frac{r}{R_p}\right)^{\lambda_{1p}-1}, \tag{69}$$

$$\sigma'_{11p} = -p \left(\frac{r}{R_p}\right)^{\lambda_{1p}-1}, \tag{70}$$

$$\sigma'_{22p} = \sigma'_{33p} = -\frac{p\xi_{5p}}{\xi_{3p}} \left(\frac{r}{R_p}\right)^{\lambda_{1p}-1}, \tag{71}$$

$$\sigma'_{12p} = -\frac{1}{s_{44p}\,\xi_{3p}} \frac{\partial p}{\partial \varphi} \left(\frac{r}{R_p}\right)^{\lambda_{1p}-1}, \tag{72}$$

$$\sigma'_{13p} = -\frac{1}{s_{44p}\,\xi_{3p}} \frac{\partial p}{\partial \nu} \left(\frac{r}{R_p}\right)^{\lambda_{1p}-1}, \tag{73}$$

where the exponent λ_{1p} and coefficients ξ_{3p}, ξ_{5p} are given by Eqs. (36) and (49), (50), respectively. The radial stress p acting at the particle-matrix boundary is given by Eqs. (92)–(95).

6.2. Cell Matrix Around Particle

With regard to Eqs. (41)–(48), (58), we get

$$u'_{1m} = -\sum_{i=1}^{2} r \left(\frac{p}{\zeta_{im}} + \frac{u_{1c0}\,\xi_{5-im}}{d\,\xi_{4+2im}}\right) \left(\frac{r}{r_c}\right)^{\lambda_{im}-1}, \tag{74}$$

$$\varepsilon'_{11m} = -\sum_{i=1}^{2} \lambda_{im} \left(\frac{p}{\zeta_{im}} + \frac{u_{1c0}\,\xi_{5-im}}{d\,\xi_{4+2im}}\right) \left(\frac{r}{r_c}\right)^{\lambda_{im}-1}, \tag{75}$$

$$\varepsilon'_{22m} = \varepsilon'_{33m} = -\sum_{i=1}^{2} \left(\frac{p}{\zeta_{im}} + \frac{u_{1c0}\,\xi_{5-im}}{d\,\xi_{4+2im}}\right) \left(\frac{r}{r_c}\right)^{\lambda_{im}-1}, \tag{76}$$

$$\begin{aligned}\varepsilon'_{12m} \quad &= -\sum_{i=1}^{2} \frac{1-\lambda_{1p}}{r_c}\left(\frac{p}{\zeta_{im}} + \frac{u_{1c0}\,\xi_{5-im}}{d\,\xi_{4+2im}}\right)\left(\frac{\partial r_c}{\partial \varphi}\right)\left(\frac{r}{r_c}\right)^{\lambda_{im}-1} \\ &-\sum_{i=1}^{2}\left[\frac{\partial}{\partial \varphi}\left(\frac{p}{\zeta_{im}}\right) + \frac{u_{1c0}}{d}\,\frac{\partial}{\partial \varphi}\left(\frac{\xi_{5-im}}{\xi_{4+2im}}\right)\right]\left(\frac{r}{r_c}\right)^{\lambda_{im}-1} \end{aligned} \tag{77}$$

$$\begin{aligned}\varepsilon'_{13m} \quad &= -\sum_{i=1}^{2} \frac{1-\lambda_{1p}}{r_c}\left(\frac{p}{\zeta_{im}} + \frac{u_{1c0}\,\xi_{5-im}}{d\,\xi_{4+2im}}\right)\left(\frac{\partial r_c}{\partial \nu}\right)\left(\frac{r}{r_c}\right)^{\lambda_{im}-1} \\ &-\sum_{i=1}^{2}\left[\frac{\partial}{\partial \nu}\left(\frac{p}{\zeta_{im}}\right) + \frac{u_{1c0}}{d}\,\frac{\partial}{\partial \nu}\left(\frac{\xi_{5-im}}{\xi_{4+2im}}\right)\right]\left(\frac{r}{r_c}\right)^{\lambda_{im}-1} \end{aligned} \tag{78}$$

$$\sigma'_{11m} = -\sum_{i=1}^{2} \xi_{2+im}\left(\frac{p}{\zeta_{im}} + \frac{u_{1c0}\,\xi_{5-im}}{d\,\xi_{4+2im}}\right)\left(\frac{r}{r_c}\right)^{\lambda_{im}-1}, \tag{79}$$

$$\sigma'_{22m} = -\sum_{i=1}^{2} \xi_{4+im}\left(\frac{p}{\zeta_{im}} + \frac{u_{1c0}\,\xi_{5-im}}{d\,\xi_{4+2im}}\right)\left(\frac{r}{r_c}\right)^{\lambda_{im}-1}, \tag{80}$$

$$\begin{aligned}\sigma'_{12m} \quad &= -\frac{1}{s_{44m}}\sum_{i=1}^{2} \frac{1-\lambda_{1p}}{r_c}\left(\frac{p}{\zeta_{im}} + \frac{u_{1c0}\,\xi_{5-im}}{d\,\xi_{4+2im}}\right)\left(\frac{\partial r_c}{\partial \varphi}\right)\left(\frac{r}{r_c}\right)^{\lambda_{im}-1} \\ &-\frac{1}{s_{44m}}\sum_{i=1}^{2}\left[\frac{\partial}{\partial \varphi}\left(\frac{p}{\zeta_{im}}\right) + \frac{u_{1c0}}{d}\,\frac{\partial}{\partial \varphi}\left(\frac{\xi_{5-im}}{\xi_{4+2im}}\right)\right]\left(\frac{r}{r_c}\right)^{\lambda_{im}-1} \end{aligned} \tag{81}$$

$$\begin{aligned}\sigma'_{13m} \quad &= -\frac{1}{s_{44m}}\sum_{i=1}^{2} \frac{1-\lambda_{1p}}{r_c}\left(\frac{p}{\zeta_{im}} + \frac{u_{1c0}\,\xi_{5-im}}{d\,\xi_{4+2im}}\right)\left(\frac{\partial r_c}{\partial \nu}\right)\left(\frac{r}{r_c}\right)^{\lambda_{im}-1} \\ &-\frac{1}{s_{44m}}\sum_{i=1}^{2}\left[\frac{\partial}{\partial \nu}\left(\frac{p}{\zeta_{im}}\right) + \frac{u_{1c0}}{d}\,\frac{\partial}{\partial \nu}\left(\frac{\xi_{5-im}}{\xi_{4+2im}}\right)\right]\left(\frac{r}{r_c}\right)^{\lambda_{im}-1} \end{aligned} \tag{82}$$

where the exponent λ_{2m} and the coefficient ξ_{2+im} $(i=1,\ldots,4)$, ζ_{jm} $(i=5,\ldots,8)$ are given by Eqs. (36) and (49), (50), (59), (60), respectively. The radial stress p acting at the particle-matrix boundary and the radial displacement u_{1c0} along the axis x_i $(i=1,2,3)$ on the cubic cell surface (i.e. for $r = d/2$) are given by Eqs. (92)–(95) and (126)–(128).

6.3. Cell Matrix Around Pore

With regard to Eqs. (41)–(48), (62), we get

$$u'_{1v} = -\frac{u_{1c0}\, r}{d}\left(\frac{r}{r_c}\right)^{\lambda_{2m}-1}, \tag{83}$$

$$\varepsilon'_{11v} = \lambda_{2m} - \frac{u_{1c0}\,\lambda_{2m}}{d}\left(\frac{r}{r_c}\right)^{\lambda_{2m}-1}, \tag{84}$$

$$\varepsilon'_{22v} = \varepsilon'_{33v} = -\frac{u_{1c0}}{d}\left(\frac{r}{r_c}\right)^{\lambda_{2m}-1}, \tag{85}$$

$$\varepsilon'_{12v} = -\frac{u_{1c0}\,(1-\lambda_{2m})}{d\, f_c}\frac{\partial f_c}{\partial \varphi}\left(\frac{r}{r_c}\right)^{\lambda_{2m}-1}, \tag{86}$$

$$\varepsilon'_{13v} = -\frac{u_{1c0}\,(1-\lambda_{2m})}{d\, f_c}\frac{\partial f_c}{\partial \nu}\left(\frac{r}{r_c}\right)^{\lambda_{2m}-1}, \tag{87}$$

$$\sigma'_{11v} = -\frac{u_{1c0}\,\xi_{4m}}{d}\left(\frac{r}{r_c}\right)^{\lambda_{2m}-1}, \tag{88}$$

$$\sigma'_{22v} = -\frac{u_{1c0}\,\zeta_{4m}}{d}\left(\frac{r}{r_c}\right)^{\lambda_{2m}-1}, \tag{89}$$

$$\sigma'_{12v} = -\frac{u_{1c0}\,(1-\lambda_{2m})}{d\, s_{44m}\, f_c}\frac{\partial f_c}{\partial \varphi}\left(\frac{r}{r_c}\right)^{\lambda_{2m}-1}, \tag{90}$$

$$\sigma'_{13v} = -\frac{u_{1c0}\,(1-\lambda_{2m})}{d\, s_{44m}\, f_c}\frac{\partial f_c}{\partial \nu}\left(\frac{r}{r_c}\right)^{\lambda_{2m}-1}, \tag{91}$$

where the exponent λ_{2m} and the coefficients ξ_{4m}, ζ_{4m} $(i = 1{,}2)$ are given by Eqs. (36) and (49), (50), respectively. The radial displacement u_{1c0} along the axis x_i $(i = 1{,}2{,}3)$ on the cubic cell surface (i.e. for $r = d/2$) is given by Eqs. (126)–(128).

6.4. Radial Stress p

With regard to Eqs. (43), (63), (67), the radial stress p acting at the particle-matrix boundary has the form

$$p = \zeta_{pm} \left(\frac{u_{1c} \zeta_{6m}}{r_c} + \beta_m - \beta_p \right) = \zeta_{pm} \left(\frac{u_{1c0} \zeta_{6m}}{d} + \beta_m - \beta_p \right), \quad (92)$$

where the coefficients ζ_{pm}, ζ_{5m}, ζ_{6m} are derived as

$$\zeta_{pm} = \frac{\xi_{3p}}{1 - \xi_{3p} \zeta_{5m}}, \quad (93)$$

$$\zeta_{5m} = \frac{\left(\frac{R_p}{r_c}\right)^{\lambda_{1m}-1} - \left(\frac{R_p}{r_c}\right)^{\lambda_{2m}-1}}{\xi_{3m} \left(\frac{R_p}{r_c}\right)^{\lambda_{1m}-1} - \xi_{4m} \left(\frac{R_p}{r_c}\right)^{\lambda_{2m}-1}}, \quad (94)$$

$$\zeta_{6m} = \frac{(\xi_{3m} + \xi_{4m}) \left(\frac{R_p}{r_c}\right)^{\lambda_{1m}-1}}{\xi_{3m} \left(\frac{R_p}{r_c}\right)^{\lambda_{1m}-\lambda_{2m}} - \xi_{4m}}. \quad (95)$$

The exponent λ_{im} and the coefficient ξ_{3p}, ξ_{2+im} $(i = 1{,}2)$ are given by Eqs. (36) and (49), respectively. The radial displacement u_{1c0} along the axis x_i $(i = 1{,}2{,}3)$ on the cubic cell surface (i.e. for $r = d/2$) is given by Eqs. (126)–(128).

6.5. Radial Displacement u_{1c0}

With regard to Eqs. (51), (55), (58), (62), the elastic energy density w_q $(q = p, m, v)$ has the form

$$w_p = \frac{r^{2\lambda_{1p}-2}}{\xi_{3p}^2} \left[\omega_{1p} \left(\frac{u_{1c0}}{d} \right)^2 + \frac{2\omega_{2p}\, u_{1c0} \, (\beta_m - \beta_p)}{d} + \omega_{3p} \, (\beta_m - \beta_p)^2 \right], \quad (96)$$

$$w_m = \omega_{1m} \left(\frac{u_{1c0}}{d} \right)^2 + \frac{\omega_{2m}\, u_{1c0} \, (\beta_m - \beta_p)}{d} + \omega_{3m} \, (\beta_m - \beta_p)^2, \quad (97)$$

$$w_v = \omega_v \left(\frac{u_{1c0}}{d} \right)^2 \left(\frac{r}{r_c} \right)^{\lambda_{2m}-2}, \quad (98)$$

where the coefficients ω_{iq} $(i=1,2,3;\ q=p,m)$, ω_v along with the coefficients ζ_{1p}, ξ_{im} $(i=7,\ldots,36)$ are derived as

$$\omega_{1p} = \zeta_{1p}\,\zeta_{6m}^2\,\zeta_{pm}^2 + \frac{1}{s_{44p}}\left\{\left[\frac{\partial\left(\zeta_{pm}\,\zeta_{6m}\right)}{\partial\varphi}\right]^2 + \left[\frac{\partial\left(\zeta_{pm}\,\zeta_{6m}\right)}{\partial\nu}\right]^2\right\}, \tag{99}$$

$$\omega_{2p} = \zeta_{1p}\,\zeta_{6m}\,\zeta_{pm}^2 + \frac{1}{s_{44p}}\left\{\frac{\partial\zeta_{pm}}{\partial\varphi}\left[\frac{\partial\left(\zeta_{pm}\,\zeta_{6m}\right)}{\partial\varphi}\right]^2 + \frac{\partial\zeta_{pm}}{\partial\nu}\left[\frac{\partial\left(\zeta_{pm}\,\zeta_{6m}\right)}{\partial\nu}\right]^2\right\}, \tag{100}$$

$$\omega_{3p} = \zeta_{1p}\,\zeta_{pm}^2 + \frac{1}{s_{44p}}\left[\left(\frac{\partial\zeta_{pm}}{\partial\varphi}\right)^2 + \left(\frac{\partial\zeta_{pm}}{\partial\nu}\right)^2\right], \tag{101}$$

$$\omega_{1m} = \sum_{i,j=1}^{2}\left(\xi_{4+i+2jm}\,\zeta_{4+i+2jm} + \frac{\zeta_{26+i+2jm}}{s_{44m}}\right)\left(\frac{R_p}{r_c}\right)^{\lambda_{im}+\lambda_{jm}-2}, \tag{102}$$

$$\omega_{2m} = \sum_{i,j=1}^{2}\left(\xi_{4+i+2jm}\,\zeta_{8+i+2jm} + \frac{\zeta_{30+i+2jm}}{s_{44m}}\right)\left(\frac{R_p}{r_c}\right)^{\lambda_{im}+\lambda_{jm}-2}, \tag{103}$$

$$\omega_{3m} = \sum_{i,j=1}^{2}\left(\xi_{4+i+2jm}\,\zeta_{12+i+2jm} + \frac{\zeta_{32+i+2jm}}{s_{44m}}\right)\left(\frac{R_p}{r_c}\right)^{\lambda_{im}+\lambda_{jm}-2}, \tag{104}$$

$$\omega_v = c_{1m} + \lambda_{2m}\left[\frac{\lambda_{2m}}{2}\left(c_{1m}+c_{2m}\right) - 2c_{2m}\right] + \left(\frac{\lambda_{2m}-1}{f_c}\right)^2\left[\left(\frac{\partial f_c}{\partial\varphi}\right)^2 + \left(\frac{\partial f_c}{\partial\nu}\right)^2\right], \tag{105}$$

$$\zeta_{1p} = \frac{\lambda_{1p}^2\left(c_{1p}+c_{2p}\right)}{2} - 2\lambda_{1p}\,c_{2p} + c_{1p}, \tag{106}$$

$$\zeta_{4+i+2jm} = \frac{\zeta_{6m}^2\,\zeta_{pm}^2}{\zeta_{im}\,\zeta_{jm}} + \frac{\xi_{5-im}\,\xi_{5-jm}}{\zeta_{2+im}\,\zeta_{2+jm}} + \zeta_{6m}\,\zeta_{pm}\left(\frac{\xi_{5-im}}{\zeta_{2+im}\,\zeta_{jm}} + \frac{\xi_{5-jm}}{\zeta_{im}\,\zeta_{2+jm}}\right),\quad i,j=1,2, \tag{107}$$

$$\zeta_{8+i+2jm} = \frac{2\,\zeta_{6m}\,\zeta_{pm}^2}{\zeta_{im}\,\zeta_{jm}} + \zeta_{pm}\left(\frac{\xi_{5-im}}{\zeta_{2+im}\,\zeta_{jm}} + \frac{\xi_{5-jm}}{\zeta_{im}\,\zeta_{2+jm}}\right),\quad i,j=1,2, \tag{108}$$

$$\zeta_{12+i+2jm} = \frac{\zeta_{pm}^2}{\zeta_{im}\,\zeta_{jm}}, \quad i,j = 1,2, \tag{109}$$

$$\zeta_{18+im} = \frac{\zeta_{6m}\,\zeta_{pm}}{\zeta_{im}} + \frac{\xi_{5-im}}{\zeta_{2+im}}, \quad i = 1,2, \tag{110}$$

$$\zeta_{19+2im} = \frac{\partial\,\zeta_{18+im}}{\partial\varphi} - \frac{d\,\zeta_{18+im}\,(\lambda_{im}-1)}{r_c}\,\frac{\partial f_c}{\partial\varphi}, \quad i = 1,2, \tag{111}$$

$$\zeta_{20+2im} = \frac{\partial}{\partial\varphi}\left(\frac{\zeta_{pm}}{\zeta_{im}}\right) - \frac{d\,\zeta_{pm}\,(\lambda_{im}-1)}{r_c\,\zeta_{im}}\,\frac{\partial f_c}{\partial\varphi}, \quad i = 1,2, \tag{112}$$

$$\zeta_{23+2im} = \frac{\partial\,\zeta_{18+im}}{\partial\nu} - \frac{d\,\zeta_{18+im}\,(\lambda_{im}-1)}{r_c}\,\frac{\partial f_c}{\partial\nu}, \quad i = 1,2, \tag{113}$$

$$\zeta_{24+2im} = \frac{\partial}{\partial\nu}\left(\frac{\zeta_{pm}}{\zeta_{im}}\right) - \frac{d\,\zeta_{pm}\,(\lambda_{im}-1)}{r_c\,\zeta_{im}}\,\frac{\partial f_c}{\partial\nu}, \quad i = 1,2, \tag{114}$$

$$\zeta_{26+2im} = \zeta_{19+2im}\,\zeta_{19+2jm} + \zeta_{23+2im}\,\zeta_{23+2jm}, \quad i = 1,2, \tag{115}$$

$$\zeta_{30+2im} = \zeta_{20+2im}\,\zeta_{19+2jm} + \zeta_{19+2im}\,\zeta_{20+2jm} + \zeta_{24+2im}\,\zeta_{23+2jm} + \zeta_{23+2im}\,\zeta_{24+2jm}, \quad i = 1,2, \tag{116}$$

$$\zeta_{32+2im} = \zeta_{20+2im}\,\zeta_{20+2jm} + \zeta_{24+2im}\,\zeta_{24+2jm}, \quad i = 1,2, \tag{117}$$

With regard to Eqs. (53), (96)–(99), the thermal-stress induced elastic energy W_q accumulated in the spherical particle ($q = p$), in the cell matrix around the spherical particle ($q = m$), and in the cell matrix around the spherical pore ($q = v$) has the form

$$W_p = \frac{8R_p^3}{\xi_{3p}^2\,(2\lambda_{1p}+1)}\left[\Omega_{1p}\left(\frac{u_{1c0}}{d}\right)^2 + \frac{2u_{1c0}\Omega_{2p}\,(\beta_m-\beta_p)}{d} + \Omega_{3p}\,(\beta_m-\beta_p)^2\right], \tag{118}$$

$$W_m = 8d\left[u_{1c0}^2\,\Omega_{1m} + u_{1c0}\,d\,\Omega_{2m}\,(\beta_m-\beta_p) + d^2\,\Omega_{3m}\,(\beta_m-\beta_p)^2\right], \tag{119}$$

$$W_v = 8\,d\,u_{1c0}^2\,\Omega_v, \tag{120}$$

where the coefficients Ω_{iq} ($i = 1,2,3$; $q = p,m$), Ω_v are derived as

$$\Omega_{ip} = \int_0^{\pi/2}\int_0^{\pi/2} \omega_{ip}\,d\varphi\,d\nu, \quad i = 1,2,3, \tag{121}$$

$$\Omega_{1m} = \int_0^{\pi/2}\int_0^{\pi/2} \sum_{i,j=1}^{2} \frac{f_c^3}{\lambda_{im} + \lambda_{jm} + 1} \left(\xi_{4+i+2jm}\, \zeta_{4+i+2jm} + \frac{\zeta_{26+i+2jm}}{s_{44m}} \right) \times \left[1 - \left(\frac{R_p}{r_c} \right)^{\lambda_{im}+\lambda_{jm}+1} \right] d\varphi\, d\nu, \tag{122}$$

$$\Omega_{2m} = \int_0^{\pi/2}\int_0^{\pi/2} \sum_{i,j=1}^{2} \frac{f_c^3}{\lambda_{im} + \lambda_{jm} + 1} \left(\xi_{4+i+2jm}\, \zeta_{8+i+2jm} + \frac{\zeta_{30+i+2jm}}{s_{44m}} \right) \times \left[1 - \left(\frac{R_p}{r_c} \right)^{\lambda_{im}+\lambda_{jm}+1} \right] d\varphi\, d\nu, \tag{123}$$

$$\Omega_{3m} = \int_0^{\pi/2}\int_0^{\pi/2} \sum_{i,j=1}^{2} \frac{f_c^3}{\lambda_{im} + \lambda_{jm} + 1} \left(\xi_{4+i+2jm}\, \zeta_{12+i+2jm} + \frac{\zeta_{32+i+2jm}}{s_{44m}} \right) \times \left[1 - \left(\frac{R_p}{r_c} \right)^{\lambda_{im}+\lambda_{jm}+1} \right] d\varphi\, d\nu, \tag{124}$$

$$\Omega_v = \int_0^{\pi/2}\int_0^{\pi/2} \frac{\omega_v f_c^3}{2\lambda_{2m} + 1} \left[1 - \left(\frac{R_v}{r_c} \right)^{2\lambda_{2m}+1} \right] d\varphi\, d\nu. \tag{125}$$

With regard to Eqs. (64), (118)–(120), the radial displacement u_{1c0} along the axis x_i ($i = 1,2,3$) on the cubic cell surface (i.e. for $r = d/2$) has the form

$$u_{1c0} = \frac{d\, \zeta_1 \left(\beta_m - \beta_p \right)}{\zeta_2}, \tag{126}$$

where the coefficients ζ_1, ζ_2 are derived as

$$\zeta_1 = \Omega_{2m} + \frac{2\Omega_{2p}}{\xi_{3p}^2 \left(2\lambda_{1p} + 1 \right)} \left(\frac{R_p}{d} \right)^3, \tag{127}$$

$$\zeta_2 = \frac{\Omega_v}{13} - 2 \left[\Omega_{1m} + \frac{\Omega_{1p}}{\xi_{3p}^2 \left(2\lambda_{1p} + 1 \right)} \left(\frac{R_p}{d} \right)^3 \right]. \tag{128}$$

7. Application to SiC-Si_3N_4 Ceramics

This section presents illustrative examples of applications of final formulae for the thermal stresses (see Sec. 6) to a real porous two-component material of the precipitate-matrix type. These illustrative examples are numerically performed for the SiC-Si_3N_4 ceramics.

Other applications of the final formulae (see Sec. 6) can be easily performed by a reader. This numerical determination can be performed by a computer software using suitable programming techniques. With regard to programming techniques within a suitable programming language (e.g. Fortran, Pascal), a derivative of the function $\psi = \psi(\eta)$ regarding the variable $\eta = \varphi, \nu$ (see e.g. Eqs. (77), (78)) can be numerically determined by the formula

$$\frac{\partial \psi}{\partial \eta} \approx \frac{\psi(\eta + \Delta\eta) - \psi(\eta)}{\Delta\eta}, \tag{129}$$

where $\psi = r_c, \xi, \zeta$, and $\eta = \varphi, \nu$. The angle steps $\Delta\varphi$, $\Delta\nu$ are required to be taken as small as possible. With regard to an experience of the author, the angle steps $\Delta\varphi = \Delta\nu = 10^{-6}$ [deg] are sufficient to be considered within the numerical determination.

Similarly, a definite integral (see Eqs. (121)–(125)) can be numerically determined by the formula

$$\Omega = \int_0^{\pi/2}\int_0^{\pi/2} \omega \, d\varphi \, d\nu \approx \sum_{j=0}^{m} \left(\sum_{i=0}^{n} \omega(i \times \Delta\varphi; j \times \Delta\nu) \, \Delta\varphi \right) \Delta\nu, \tag{130}$$

where n, m are integral parts of the real numbers $\pi/(2\,\Delta\varphi)$, $\pi/(2\,\Delta\nu)$, respectively. With regard to an experience of the author, the angle steps $\Delta\varphi = \Delta\nu = 0.01 - 0.1$ [deg] are sufficient to be considered within the numerical determination.

Considering the same conditions concerning the angle steps $\Delta\varphi$, $\Delta\nu$ as related to Eq. (130), the average value $\overline{\kappa}$ of the function $\kappa = \kappa(\varphi, \nu)$ of the variables $\varphi, \nu \in \langle 0, \pi/2 \rangle$ can be numerically determined by the formula

$$\overline{\kappa} = \left(\frac{2}{\pi}\right)^2 \int_0^{\pi/2}\int_0^{\pi/2} \kappa \, d\varphi \, d\nu \approx \left(\frac{2}{\pi}\right)^2 \sum_{j=0}^{m} \left(\sum_{i=0}^{n} \kappa(i \times \Delta\varphi; j \times \Delta\nu) \, \Delta\varphi \right) \Delta\nu, \tag{131}$$

where $\kappa = p, u_{1c0}, (\sigma'_{11v})_{r=r_c}$.

Table 1 presents material parameters of the SiC particle and the Si_3N_4 matrix of the SiC-Si_3N_4 ceramics. The SiC-Si_3N_4 ceramics represents a porous two-component material of a precipitate-matrix type. Pores in the Si_3N_4 matrix are a consequence of powder metallurgy processes [9]. The relaxation temperature $T_r = 665°C$ is determined by the formula $T_r = 0.35\,T_m$ [9], where $T_m = T_{mm}$ is melting temperature of the SiC-Si_3N_4 ceramics.

Table 1. Material parameters of the SiC-Si_3N_4 ceramics [9].

	Particle ($q = p$) **SiC**	**Matrix** ($q = m$) **Si_3N_4**
E_q [GPa]	360	310
μ_q	0.19	0.235
α_q [$10^{-6}K^{-1}$]	4.15	2.35
T_{mq} [°C]	2730	1900

As shown in Fig. 5, the dependence $\overline{p} - v_p$ (see Eq. (131)) for the radial stress $p = p(\varphi, \nu)$ acting at the SiC-Si_3N_4 boundary exhibits a decreasing course for $v_p \in (0, v_{pmax})$ (see Eq. (2)) in contrast to an increasing course of the radial displacement $u_{1c0} = u_{1c0}(v)$ (see Fig. 6) along the axis x_i ($i = 1,2,3$) on the cubic cell surface (i.e. for $r = d/2$) (see Eqs. (126)–(128)), where $R_p = 3\ \mu$m, $R_v = 0.1\ \mu$m, $T_r = 665°C$, $T_f = 20°C$.

With regard to $\beta_p > \beta_m$ (see Eq. (1), Tab. 1), the dimension d of the cubic cell with the spherical particle is changed to $d - u_{1c)}$ due to the thermal stresses. Consequently, the dimension d of the cubic cell with the spherical pore is changed to $d + u_{1c)}$ due to the thermal stresses.

Finally, the increasing dependence $\left(\overline{\sigma'_{11v}}\right)_{r=r_c} - v_p$ (see Eq. (131)) for the radial stress $(\sigma'_{11v})_{r=r_c} = p(\varphi, \nu)$ acting on a surface of the cubic cell with the spherical pore (i.e. for $r = r_c$, see Eqs. (5)–(10)) is shown in Fig. 7

Conclusion

Results of this chapter are as follows.

The analytical model of thermal stresses in a porous multi-particle-matrix system with isotropic components is determined (see Secs. 2, 6). The analytical model results from fundamental equations of mechanics of an elastic solid continuum (see Sec. 3).

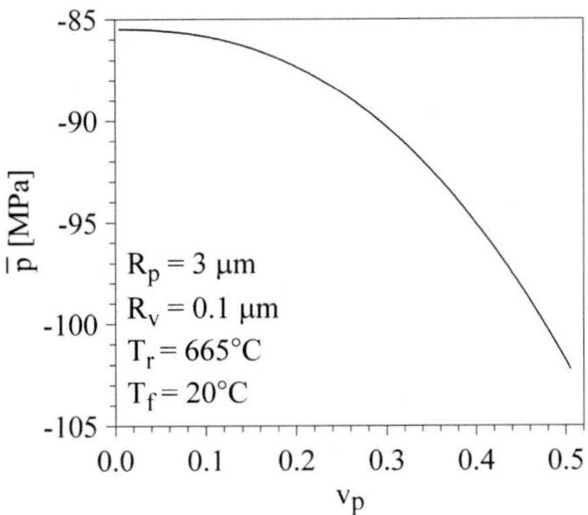

Figure 5. The function $\overline{p} = \overline{p}\,(v_p)$ of the volume fraction $v_p \in (0, v_{pmax}\rangle$ (see Eqs. (2), (131)) of the SiC particle, where $R_p = 3$ μm, $R_v = 0.1$ μm, $T_f = 20$°C, $T_r = 665$°C. The radial stress $p = p\,(\varphi, \nu, v)$ (see Eqs. (92)–(95)) acts at the SiC-Si_3N_4 boundary.

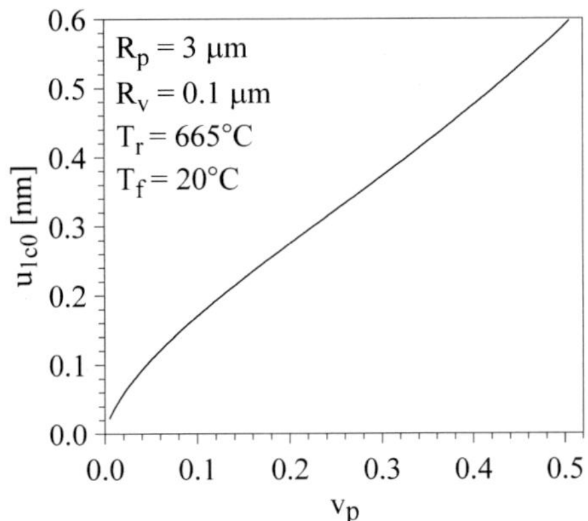

Figure 6. The radial displacement $u_{1c0} = u_{1c0}\,(v_p)$ along the axis x_i ($i = 1,2,3$) on a surface of the cubic cell with the spherical pore (i.e. for $r = d/2$) (see Eqs. (126)–(128)) as a function of the volume fraction $v_p \in (0, v_{pmax}\rangle$ (see Eq. (2)) of the SiC particle, where $R_p = 3$ μm, $R_v = 0.1$ μm, $T_f = 20$°C, $T_r = 665$°C.

The porous multi-particle-matrix system consists of isotropic spherical particles and spherical pores which are both periodically distributed in an isotropic infinite matrix (see Fig. 1). The thermal stresses are a consequence of different

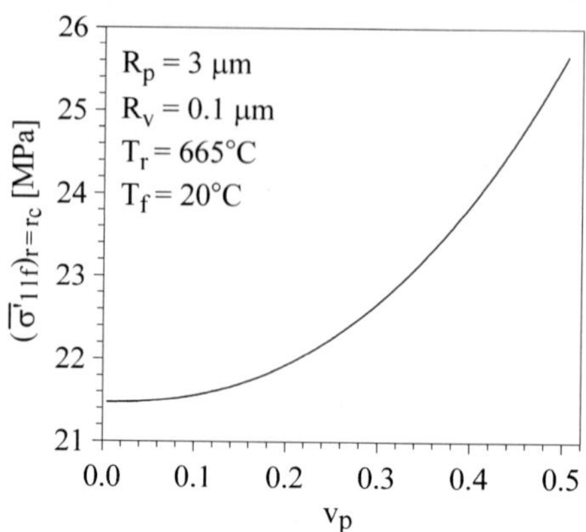

Figure 7. The dependence $\left(\overline{\sigma'_{11v}}\right)_{r=r_c} - v_p$ on the volume fraction $v_p \in (0, v_{pmax}\rangle$ (see Eqs. (2), (131)) for the radial stress $(\sigma'_{11v})_{r=r_c} = p(\varphi, \nu)$ acting on a surface of the cubic cell with the spherical pore (i.e. for $r = r_c$, see Eqs. (5)–(10), (88)), where $R_p = 3\ \mu$m, $R_v = 0.1\ \mu$m, $T_f = 20$°C, $T_r = 665$°C.

thermal expansion coefficients of the isotropic matrix and isotropic particle (see Eq. (1)).

The infinite matrix is imaginarily divided into identical cubic cells (see Fig. 1), and the thermal stresses are thus investigated within the cubic cell. Each cell contains either a central particle or a central pore (see Sec. 2). This porous multi-particle-matrix system represents a model system which is applicable to porous two-component materials of a precipitate-matrix type. Radii and volume fractions of both the particles and of the pores along with the cubic cell dimension represent microstructural parameters of the porous two-component material. The thermal stresses are then functions of these microstructural parameters. Numerical values for a real porous two-component material of the precipitate-matrix type (SiC-Si_3N_4 ceramics) are obtained.

Acknowledgments

This work was supported by the Slovak Grant Agency under the contract VEGA 2/0081/16, 2/0120/15, 2/0113/16 and by the Slovak Research and Development Agency under the contract APVV-0147-11.

References

[1] Greisel, M.; Jäger, J.; Moosburger-Will, J.; Sause, M.G.R.; Mueller, W.M.; Horn, S. *Composites: Part A* **2014,** *66,* 117-127.

[2] Meijer, G.; Ellyin, F.; Xia, Z. *Composites: Part B* **2000,** *31,* 29–37.

[3] Węglewski, W.; Basista, M.; Manescu, A.; Chmielewski, M.; Pietrzak, K.; and Schubert, Th. *Composites: Part B* **2014,** *67,* 119-124.

[4] Nabavi, S.M.; Ghajar, R. *Int. J. Eng. Sci.* **2010,** *48,* 1811-1823.

[5] Lee, J.; Hwang, K.-Y. *Int. J. Eng. Sci.* **1996,** *34,* 901–922.

[6] Tanigawa, Y.; Osako, M.K. *Int. J. Eng. Sci.* **1986,** *24,* 309–321.

[7] Abedian, A.; Szyszkowski, W.; Yannacopoulos, S. *Compos. Sci. Technol.* **1999,** *59,* 41–54.

[8] Ceniga, L. *Analytical Model of Thermal Stresses in Composite Materials I;* Nova Science Publishers: New York, US, 2008, pp 66–67, 136–137.

[9] Skočovský, P.; Bokůvka, O.; Palček, P. *Materials Science;* EDIS: Žilina, SK, 1996, pp 35–48.

[10] Selsing, J. *J. Amer. Cer. Soc.* **1961,** *44,* 419–419.

[11] Davidge, R.W.; Green, T.J. *J. Mater. Sci.* **1968,** *3,* 629–634.

[12] Mastelaro, V.R.; Zanotto, E.D. *J. Non-Crystal. Sol.* **1996,** *194,* 297–304.

[13] Mastelaro, V.R.; Zanotto, E.D. *J. Non-Crystal. Sol.* **1999,** *247,* 79–86.

[14] Diko, P. *Supercond. Sci. Technol.* **1998,** *11,* 68–72.

[15] Diko, P.; Fuchs, G.; Krabbes, G. *Physica C* **2001,** *363,* 60–66.

[16] Diko, P. *Supercond. Sci. Technol.* **2004,** *17,* R45–R58.

[17] Diko, P. *Int. J. Mater. Product Technol.* **2014,** *49,* 97-128.

[18] Chmelík, F.; Trník, A.; Štubňa, I.;, Pešička, J. *J. Eur. Ceram. Soc.* **2011,** *31,* 2205-2209.

[19] Serbena, F.C.; Zanotto, E.D. *J. Non-Crystal. Sol.* **2012,** *358,* 975–984.

[20] Bidulský, R.; Bidulská, J.; Grande, M.A. *Arch. Metallur. Mater.* **2013,** *58,* 365–370.

[21] Kvačkaj, T.; Kočisko, R.; Bidulský, R.; Bidulská, J.; Bella, P.; Lupták, M.; Kováčová, A.; Bacso, J. *Mater. Sci. For.* **2014,** *782,* 379–383.

[22] Bidulský, R.; Bidulská, J.; De Oro, R.; Hryha, E.; Maccarini, M.; Forno, I.; Grande, M.A. *Act. Phys. Polon. A* **2015,** *128,* 647–650.

[23] Knapek, M.; Húlan, T.; Minárik, P.; Dobroň, P.; Štubňa, I.; Strásk‌á, J.; Chmelík, F. *J. Eur. Ceram. Soc.* **2016,** *36,* 221-226.

[24] Mizutani, T. *J. Mater. Sci.* **1996,** *11,* 483–494.

[25] Li, Sh.; Sauer, R.A.; Wang, G. *J. Appl. Mech.* **2007,** *74,* 770–783.

[26] Li, Sh.; Sauer, R.A.; Wang, G. *J. Appl. Mech.* **2007,** *74,* 784–797.

[27] Kushch, V.I. *Prikladnaja Mekhanika* **1985,** *21,* 18-27.

[28] Kushch, V.I. *Int. Appl. Mech.* **2004,** *40,* 893-899.

[29] Kushch, V.I. *Int. Appl. Mech.* **2004,** *40,* 1042-1049.

[30] Wu, Y.; Dong, Z. *Mater. Sci. Eng.* **1995,** *203,* 314–323.

[31] Sangani, A.S.; Mo, G. *J. Mech. Phys. Sol.* **1997,** *45,* 2001–2031.

[32] Mura, T. *Micromechanics of Defects in Solids;* Martinus Nijhoff Publishers: Dordrecht, NL, 1987, pp 1–3.

[33] Ceniga, L. *Analytical Models of Thermal Stresses in Composite Materials IV;* Nova Science Publishers: New York, US, 2015, pp 34–37.

[34] Brdička, M.; Samek, L.; Sopko, B. *Mechanics of Continuum;* Academia: Prague, CZ, 2000, pp 78–83.

[35] Trebuňa, F.; Šimčák, F.; Jurica, V. *Elasticity and Strength I;* Technical University: Košice, SK, 2005, pp 57–63.

[36] Trebuňa, F,;, Šimčák, F.; Jurica, V. *Examples and Problems of Elasticity and Strength I;* Technical University: Košice, SK, 2002, pp 134–138.

[37] Rektorys, K. *Review of Applied Mathematics;* SNTL: Prague, CZ, 1973, pp 253–267.

In: Ceramic Materials
Editor: Jacqueline Perez
ISBN: 978-1-63485-965-3

Chapter 5

AN UPDATE OF CURRENT KNOWLEDGE ON ADHESION TO ZIRCONIA DENTAL RESTORATIONS

***M. L. P. D. Engler*[1], *E. Ruales*[1], *M. Özcan*[2] *and C. A. M. Volpato*[1,*]**

[1]Department of Dentistry, Federal University of Santa Catarina, Florianópolis, Brazil

[2]Dental Materials Unit, Center for Dental and Oral Medicine, Clinic for Fixed and Removable Prosthodontics and Dental Materials Science, University of Zurich, Zurich, Switzerland

Abstract

Yttria-stabilized tetragonal zirconia polycrystal (Y-TZP) has been used in dentistry in order to manufacture prosthetic frameworks, monolithic crowns and implant abutments due to its superior mechanical properties, biocompatibility, chemical stability and appropriate aesthetics as opposed to other materials. The survival of dental ceramic restorations depends on durable bond strength between the restorative material, composite resin luting cement and the tooth surface. However, it is difficult to establish a durable mechanical or chemical adhesion in zirconia-based prostheses since yttrium-stabilized zirconia is an oxide ceramic that does not contain silicon dioxide (SiO_2) phase in its microstructure. In order to achieve strong and reliable adhesion between resin composite luting cements

[*] E-mail address: claudia.m.volpato@ufsc.br (Corresponding author).

and zirconia surfaces, it is crucial to employ a method that does not impair the mechanical properties and at the same time rendering it compatible with the luting cement. Furthermore, the chosen method should be practical, easy to perform and should not cause *t*→*m* phase transformation. Several methods and protocols for conditioning zirconia surfaces prior to adhesive cementation have been suggested in the literature such as physical, physicochemical and chemical methods. Typically, while physical surface conditioning methods are based on employing air-borne particle abrasion with alumina particles, physicochemical methods use silica-coated alumina particles followed by silanization. It is also possible to activate the zirconia surface chemically using functional-monomer containing adhesive promoters in the form of adhesive cements or primers. Generally, combination of micromechanical and chemical surface conditioning methods are preferred to enhance adhesion to zirconia. As this ceramic demonstrates superior properties compared to other ceramics, it is essential to study the peculiar characteristics of dental zirconia after surface conditioning methods and suggest one that does not damage its favorable mechanical properties. This chapter will provide the recent information on the use of zirconia in dentistry, its characteristics and indications, with a particular emphasis on surface conditioning methods to promote adhesion of resin-based materials to zirconia.

Keywords: zirconia, adhesion, prosthetic restorations, dentistry

1. Introduction

The science of ceramic materials is evolving, including the development and introduction of new ceramic materials with high mechanical strength that allows for new indications and applications in dentistry. During the last few decades, new materials have been introduced in dentistry that could be used in conjunction with CAD/CAM technologies in order to solve dental problems that used to be limited to metal-ceramic restorations (Özcan & Volpato, 2015). Among many other ceramics, zirconia (ZrO_2) received more attention of dentists and researchers due to the its favourable mechanical properties (Miyazaki et al., 2013; Ferrari, Vichi & Zarone, 2015), offering long term durability (Piascik et al., 2009) and versatility of aplications such as fixed dental prosthesis (FDPs), root posts, orthodontic brackets, implants and implant abutments (Özcan & Bernasconi, 2015).

1.1. Main Caracteristics of Zirconia

Zirconium dioxide (hereafter: zirconia) is an allotrope meaning that it can be found in more than one crystalline phase with the same chemical composition, namely monoclinic, tetragonal and cubic phases (Curtis, Wright & Fleming, 2006). However, in order to maintain zirconia in the tetragonal state at low temperatures (below 1770°C to room temperature), it is necessary to use a stabilizer such as yttrium, magnesium, cerium and calcium oxides, all of which could lead zirconia to a metastable state (Guan, Zhang & Liu, 2015). Yttria-stabilized tetragonal zirconia polycrystal is composed of ZrO_2 crystals and contains 2.5 to 3.5% yttrium oxide (Y_2O_3) (Vagkopoulou et al., 2009). As a biomaterial in medicine, zirconia stabilized by yttrium oxide is the only material with an ISO standard (13356.2008) (International Standards Organization, 2008).

Because of its metastable state, the Y-TZP can undergo a process called "transformation toughening". This phenomenon is a phase transformation induced by external forces. The impact of a magnitude of force could create a crack on the surface of the material but the metastable particles convert from tetragonal to monoclinic phase resulting in volumetric expansion of approximately 5% (Lughi & Sergo, 2010). This peculiar characteristic increases mechanical properties of zirconia (Passos et al., 2014). YTZP has fracture strength of >1000 MPa and toughness of 5-7 MPa m0.5, being the highest reported for a dental ceramic (Castro et al., 2012).

Apart from mechanical properties, its aesthetic properties, biocompatibility, chemical stability and low thermal conductivity are determinant factors in order to choose this ceramic for several dental applications (Sciasci et al., 2015). Y-TZP is a chemically inert material and no local or systemic adverse effects have been reported. When it comes to cytotoxic, oncogenic or mutagenic properties, no adverse effects were observed on neither on fibroblasts nor on blood cells (Vagkopoulou et al., 2009). In vitro tests showed that zirconia ceramics have similar cytotoxicity levels with alumina (Piconi & Maccauro, 1999), and the in vivo behavior of zirconia compared with alumina did not show differences as regards to tissue reaction (Garvie & Nicholson, 1972). From optical point of view, being white, it is considered as an aesthetic material. Despite its high refractive index, it is still more translucent than metal (Vagkopoulou et al., 2009). With the purpose of enhancing aesthetics, some manufacturers offer coloured zirconia frameworks but limited information is available on the long-term colour stability (Reich & Hornberger, 2002). Recently, high-translucent zirconia has been introduced that contains smaller particles in the microstructure, having higher density and cubic zirconia in the composition, making a hybrid of tetragonal-cubic zirconia (Zhang & Kim, 2009).

2. Adhesive Bonding for Zirconia

Adhesive bonding of zirconia on dental tissues using resin based luting cements were recommended for improved retention (Blatz, Sadan & Kern, 2003), marginal adaptation (Derand et al., 2005) and the possibility of achieving more conservative dental restorations. Also, longevity of ceramic restorations is highly dictated by the adhesive bonding procedures. Likewise, long-term success is highly affected by the surface conditioning methods employed on the intaglio surfaces of zirconia-based reconstructions (Senyilmaz et al., 2007; Blatz, Sadan & Kern, 2003; Ozcan, Nijhuis & Valandro, 2008).

Several luting cements have been proposed to be used with zirconia reconstructions, including glass ionomer (Blatz et al., 2007), conventional composite resin and self-etching cements (Attia, 2011). Nonetheless, adhesion between resin-based materials and zirconia is compromised as zirconia is an oxide ceramic without silica phase in its composition that is also acid-resistant (Kern & Wegner, 1998; Derand & Derand, 2000; Blatz, Sadan & Kern, 2003; Thompson et al., 2011). Hence, it is not possible to condition the surface of this material with hydrofluoric acid etching to create micromechanical retention (Della Bona et al. 2007).

On the contrary, in silica based ceramics the use of hydrofluoric acid removes the vitreous matrix, increasing the surface roughness and energy in order to achieve a micromechanical bonding with resin-based luting cement. Also, in silica-based ceramics, the use of silane coupling agents allows the union of silicon dioxide (SiO_2) present in the ceramic surface with the organic matrix of the resin composite luting cements (Mattiello et al., 2013). In the absence of SiO_2 on the surface of the zirconia, silane coupling agents are ineffective in adhesive bonding procedures (Tsuo, Atsuta & Yoshida, 2006). Additionally, zirconia is an inert substrate with low surface energy and wettability (Özcan & Vallitu, 2003). It is virtually a flat material, free of roughness without a pattern of detectable microretention as it is composed of very small crystals (Janda et al., 2003) with minimum space between particles (Duran & Moure, 1984).

2.1. Surface Conditioning Methods for Zirconia Surfaces

Since zirconia surface does not favour durable adhesive bonding, numerous surface conditioning methods have been proposed in order to improve the interaction between resin-based luting cements and the zirconia surface (Sciasci et al., 2015; Burke et al., 2002; Derand, Molin & Kvam, 2005; Subaşı & İnan, 2014). Among all methods proposed, the simplest, functional and widely employed method used is air-borne particle abrasion using Al_2O_3 particles. In this method, particles ranging from 25 to 120 μm have been indicated while shear bond strength values reported statistically similar results independent of the particle size (Sciasci et al., 2015). Also, when bigger particles are used, possible ditching between the resin composite luting cement and zirconia surface could occur (Özcan, Nijhuis & Valandro, 2008). For this reason, particles of 50 μm are commonly accepted for air-abrasion protocols used for zirconia (Özcan, Nijhuis & Valandro, 2008; Blatz et al., 2010).

Air-abrasion is typically performed using an intraoral air-abrasion device at a pressure between 2.3 and 3 bar for 12 to 15 seconds, maintaining a distance of 10 mm from the zirconia surface (Blatz et al., 2010; Özcan, Nijhuis & Valandro, 2008; Michida et al., 2015). This type of mechanical conditioning method could also remove contamination from cementation surfaces, clean the desired area and produce a rough surface on zirconia that increase the surface area and improve its surface energy (Özcan, Nijhuis & Valandro, 2008; Souza et al., 2013). Despite such advantages of air-abrasion protocols, air-particle abrasion without the use of primers could result in initial high bond strength to zirconia surface but the bond decreases drastically after long-term aging conditions (Kern, Barloi & Yang, 2009; Özcan et al., 2013).

One other suggested physicochemical conditioning method is tribochemical silica coating, a procedure usually applied in dental laboratories using Al_2O_3 particles coated with silica layer (30 - 110 μm) (Burke et al., 2002; Sarmento et al., 2014). This conditoning method not only creates micro-roughness on the surface but also chemically activates the surface to react with the silane coupling agents (Özcan, Nijhuis & Valandro, 2008; Atsu et al., 2006; Kern & Wegner, 1998). The silanol groups (Si-OH) and later their reaction with the silica surface form siloxane (-Si-O-Si-O-) network (Özcan, Nijhuis & Valandro, 2008) that is capable of achieving covalent bonds at resin cement/silica-coated ceramic interface (Matinlinna et al., 2006; Valandro et al., 2007; Xible et al., 2006), justifying the use of this conditioning method.. Yet, poor silicon surface could suffer from after hydrothermal aging and impair the achieved adhesion

(Matinlinna et al., 2006). Therefore, tribochemical silica coating with the use of appropriate primers seems mandatory (Kern, Barloi & Yang, 2009).

The optimum protocol for silica coating of zirconia surface was described as follows: (Özcan, 2013; Özcan, 2014)

1. Check the fit of the restoration using silicone based materials.
2. Mark the area to be air-abraded with a pencil.
3. Protect the outer surface of the restoration utilizing glycerin gel.
4. Air abrade the cementation surface with silica coated alumina particles or silica particles only, using a chairside air-abrasion device. The particles size should range between 30 to 50μm at a pressure of 0.5 to 2.5 bar for 20 seconds, maintaining a distance of 10 mm from zirconia surface in circling motions.
5. Ultrasonic cleaning of the restoration in ethanol for 10 min.

After this surface conditioning protocol, the intaglio surfaces should be able to interact with silane coupling agents and thereafter with the resin-based luting cements.

Current evidence indicates superior performance of tribochemical silica coating in improving adhesion to zirconia when compared to air-abrasion with alumina (Özcan & Vallitu 2003; Atsu et al. 2006; El-Korashy & El-Refai, 2014; Erdem et al., 2014; Özcan & Bernasconi 2015).

Despite the fact that the most widely used methods to improve adhesion to zirconia is air-abrasion with alumina and tribochemical silica coating, other conditioning methods are also being studied to induce increased surface bond area, surface energy and wettability (Mattiello et al., 2013; Kim et al., 2011). One such method is plasma coating with hexamethyldisiloxane, a rapid process performed at low temperatures by a reactor that generates an ionized gas when deposited on the zirconia surface. Although in part the reaction is explained by covalent bonds, adhesion mechanism of this surface modification remains unclear (Derand, Molin & Kvam, 2005). The application of micropearls of low fusing porcelain or vapor deposition of silicone tetrachloride ($SiCl_4$) has also been investigated to obtain a silica layer on the zirconia surface (Derand, Molin & Kvam, 2005; Cura et al., 2012; Castro et al., 2012; Derand et al., 2008). This method have shown improved bond strength of resin composite luting cements but glass deposition could interfere with the internal adaptation of restorations.

Other mechanical conditioning methods contemplate the use of erbium-doped yttrium aluminum garnet (Er:YAG) laser to modify zirconia through the removal of particles by micro explosions and vaporization, in a process named ablation

(Cavalcanti et al., 2009). Er:YAG laser appears to have minimal impact on surface modifications as opposed to air-abrasion protocols that is attributed to lower laser energy absorption due to the fact that it is a water-free material (Subaşı & İnan, 2014). This procedure might damage the surface of zirconia due to severe temperature modifications that could be responsible for phase transformation (Guazzato et al., 2005). On the other hand, CO_2 laser showed improved adhesion as a result of microcracks created on the zirconia surface allowing resin penetration. Unfortunately, this could yield to subsurface damage of zirconia that eventually weakens the material (Kasraei et al., 2014). Nevertheless, results are still contradictory for improving bond strength of resin materials to zirconia with the use of laser technologies (Subaşı & İnan, 2014; Mattiello et al., 2013).

One other interesting method in this respect is selective infiltration etching (SIE) that transforms the relatively smooth zirconia surface into a 3-dimensional network of inter-grain porosity, making it a highly retentive surface (Aboushelib et al., 2009). Principally, heat-induced maturation and grain boundary diffusion that consist a low temperature fused glass ceramic layer is selectively applied to the surface and subsequently etched with 5% hydrofluoric acid (Aboushelib et al., 2009; Aboushelib, 2011). Despite the increased initial bond strength, a stable bond promoted with this method after long-term aging is still questionable (Mattiello et al., 2013; Cheung & Botelho, 2015).

Other attempts to enhance adhesion of resin materials to zirconia involve the use of acidic agents (Liu et al. 2015)), namely etching zirconia with an acidic solution at 100ºC for 25 minutes or etching with hydrofluoric acid (48%) at 100ºC for 25 minutes. After both methods, the specimens were sintered following the manufacturer`s protocol. With such acid etching protocols it is claimed that zirconia grains are removed or dislodged, creating a rough surface that could result in an enhanced resin to zirconia interlocking.

3. Composite Resin Cements

Adhesion to zirconia could also be improved with the use of resin cements containing functional monomers due to their special affinity to metal oxides (Kern & Wegner, 1998; Blatz et al., 2010; Passos et al., 2010; Mattiello et al., 2013). Phosphate ester monomers in such resin cements, such as 10-methacryloyloxydecyl-dihydrogenphosphate (MDP) could bond directly to metal oxides (Kern & Wegner, 1998). Thus, the presence of hydroxyl groups on the zirconia surface could activate the reaction with the phosphate ester monomers of

MDP (Oyagüe et al., 2009; Özcan & Bernasconi, 2015). The use of such monomers is responsible for water-resistant chemical bond with zirconia that is significantly higher compared to the bond strength with conventional bis-GMA resin cements. This chemical bond seems stable even after aging in water storage or the decrease of bond strength is not statistically significant (Kern & Wegner, 1998). The use of MDP based resin cements is indispensable for achieving proper adhesion to zirconia surface (Kern & Wegner, 1998; Blatz et al., 2010; Oyagüe et al., 2009). The bond strength could even be further improved with the use of MDP-containing silane coupling agents after silica-coating (Özcan & Bernasconi, 2015) (Figure 1).

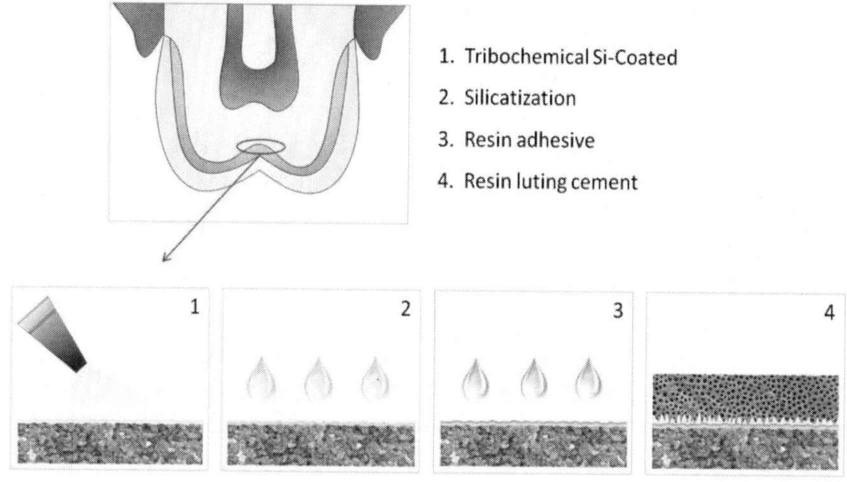

Figure 1. 1-4. 1) Tribochemical silica coating of the zirconia surface; 2) Silane coupling agent application; 3) Resin luting cement application; 4) Cemented zirconia restoration.

Since application of surface conditioning methods are rather cumbersome and require additional armamentarium in the clinical practice, self-adhesive resin cements are introduced that require no initial surface conditioning of zirconia or the dental tissues with the advantages of simplified clinical steps, low incidence of postoperative sensitivity (Blatz et al, 2010). Manufacturers promote the use of self-adhesive resin cements as a chemical conditioning method for zirconia, dismissing any other mechanical or chemical conditioning methods that would be mandatory with conventional resin cements (Özcan & Volpato, 2015). Such resin cements show adequate adhesion to different substrates, including zirconia (Passos et al., 2010; Sciasci et al., 2015; Ehlers et al., 2015). The presence of functional acidic monomers in their composition allows for chemical bond with the metal oxides on the zirconia surface (Blatz et al., 2010). Nevertheless, limitations of self-adhesive resin cements are referred as low degree of conversion, low diffusion level into dentin and less hydrolytic stability (Monticelli et al., 2008; Vrochari et al., 2009; Özcan & Bernasconi, 2015). A comparison ceramics indicated for frameworks, their different characteristics and the corresponding surface conditioning methods are summarized in Table 1.

Table 1. Overview of ceramics indicated for frameworks, their different characteristics and the corresponding surface conditioning methods

Type of ceramic	Indication	Limitation	Flexural resistance	Hydrofluoridric acid sensivity	Surface treatments
Lithium dissilicate	Single crowns, anterior and posterior FDP's up 3 unit	Not indicated for extended prostheses	350MPa	Acid sensitive	- Hidrofluoridric acid etching - Silane application
Alumina	Ceramic infrastructures up to 4 units	Not indicated for areas with high aesthetic demand involving translucently	600MPa	Acid resistance	- Silica coating or air-particle abrasion with aluminium oxide - Silane application
Zirconia	Anterior and posterior FDPs, implant abutments and root posts	Not indicated for areas with high aesthetic demand involving translucently	1000MPa	Acid resistance	-Air-particle abrasion with aluminum oxide -Tribochemical silica coating - Hexamethyldisiloxane plasma spray - Micropearls of low fusing porcelain application - Silicon tetrachloride vapor deposition - Er:YAG laser CO_2 laser - Heated HF 48% - Heated acid solution

3.1. Adhesive Interface Stability

Based on the currently available data in the literature, some concerns are present regarding the zirconia-resin adhesive interface in that the process of aging conditions, simulating the oral environment, decreases the hydrolytic stability or the mechanical resistance of zirconia (Tzanakakis et al., 2016; Lung & Matinlinna, 2012). The main reasons for approximately 50% of reduction from baseline to aged conditions are the hydrolytic effect of water and the moisture of the environment (Abousheilib et al., 2009).

Typically, thermocycling aging method is used to simulate the worst-case oral environment and aging of adhesive interfaces where water at alternating temperatures between 5 and 55°C causes thermal expansion and contraction of the bonded materials, aging the bonded interface (Liu et al., 2014). In fact, the increase in surface area with the air-particle abrasion with different particles (silica, alumina, zirconia) for adhesion can help to improve the resistance against the hydrolytic effects on the resin-zirconia adhesive interface (Liu et al., 2014; Liu et al., 2015; Tzanakakis et al., 2016). However, even with silica coating, although the initial bond strength is doubled compared to the non-conditioned control group, after aging conditions the mean bond strength also decrease. This was partially attributed to the loosening of the attached silica and the hydrolytic degradation of the siloxane interface layer (Liu et al. 2013).

Unfortunately, findings regarding this subject are difficult to compare between the studies since different zirconia materials, experimental tests, size of specimens and aging conditions greatly affect the results (Özcan & Bernasconi, 2015). However, at least it can be stated that there is practically no adhesion on zirconia when no surface conditioning method is employed, providing that the achieved results after all types conditioning methods tend to show poor long-term stability (Tzanakakis et al., 2016).

4. Phase Transformation

The possibility to optimize the adhesion between resin luting composite and zirconia with surface conditioning methods is essential. However, especially mechanical surface conditioning methods could alter the zirconia microstructure (Karacoca & Yilmaz, 2009; Thompson et al., 2011). In this regard, the higher the amounts of monoclinic phase in zirconia, the higher the chances for failure. Accordingly, air-abrasion methods increase the monoclinic content in the zirconia microstructure (Kosmac et al., 1999) since the particle organization changes and fragile areas are created on the surface of the material. It has to be however noted that air-abrasion could be performed using different particle sizes (e.g., 30 – 250 μm) and deposition duration (e.g., 5, 15 and 30 seconds) that have direct affect on the amount of monoclinic content (Turp et al. 2013). The results of XRD analysis demonstrate higher amounts of monoclinic phase with the increase in the particle size and durations. There is thus concern on the possible impact of air-abrasion protocols that could create slow crack growth of zirconia, producing surface defects, reduce the mechanical strength and consequently affecting its long-term performance (Zhang et al. 2004; Zhang et al., 2006).

Interestingly however, air-particle abrasion with 110 μm particle size can create a thin compressive layer on the zirconia surface and even increase the strength of zirconia as it initiates transformation when the material is subjected to stresses (Kosmac et al., 1999; Kosmac et al., 2000). Yet, this compressive layer and the amount of monoclinic phase content are highly dictated by the deposition duration and pressure (Turp et al., 2013).

Since surface conditioning methods commonly introduce defects on the zirconia microstructure, fused glass infiltration method was suggested on the top and bottom surface of zirconia (Zhang & Kim, 2009). This process resulted in a graded glass/ceramic/glass material with an outer surface having residual glass layer, a dense ceramic core in the middle and a graded glass ceramic layer internally. These characteristics are claimed to improve optical properties and overcome adhesion related problems. In this graded material, glass content is relatively high at the residual glass layer interfaces and gradually transforms to a dense Y-TZP in the interior regions. With this architecture, a network of glass-coated zirconia grains are exposed that serve for ideal surface morphology for silanization and adhesive resin penetration. The graded material approach needs to be etched with hydrofluoric acid and silanized that could eventually eliminate the need for surface conditioning using air-abrasion technologies required for pure Y-TZP (Zhang & Kim, 2009). However, studies in this regard are limited.

Conclusion

A lack of consensus exists on the unique method to bond resin-based luting cements to zirconia resulting in confusion in the dental community. Yet, the available evidence is in favour of micromechanical retention achieved through air-abrasion protocols, especially using tribochemical silica coating, followed by MDP containing silane coupling agents and MDP based resin luting cements.

References

Aboushelib, M.N. (2011). Evaluation of zirconia/resin bond strength and interface quality using a new technique. *J. Adhes. Dent.* 13, 255-60.

Aboushelib, M.N., Mirmohamadi, H., Matinlinna, J.P., Kukk, E., Ounsi, H.F., Salameh, Z. (2009). Innovations in bonding to zirconia-based materials. Part II: Focusing on chemical interactions. *Dent.Mater.* 25, 989-93.

Atsu, S.S., Kilicarslan, M.A., Kucukesmen, H.C., Aka, P.S. (2006). Effect of zirconium-oxide ceramic surface treatments on the bond strength to adhesive resin. *J. Prosthet. Dent.* 95, 430-6.

Attia, A. (2011). Bond strength of three luting agents to zirconia ceramic-influence of surface treatment and thermocycling. *J. Appl. Oral Sci.*19, 388-395.

Blatz, M.B., Chiche, G., Holst, S., Sadan, A. (2007). Influence of surface treatment and simulated aging on bond strengths of luting agents to zirconia. *Quintessence Int.* 38, 745-753.

Blatz, M.B., Phark, J.H., Ozer, F., Mante, F.K., Saleh, N., Bergler, M., Sadan, A. (2010). In vitro comparative bond strength of contemporary self-adhesive resin cements to zirconium oxide ceramic with and without air-particle abrasion. *Clin. Oral. Investig.* 14, 187-92.

Blatz, M.B., Sadan, A., Kern, M. (2003) Resin-ceramic bonding: a review of the literature. *J. Prosthet. Dent.* 89, 268-274.

Burke, F.J., Flemming, G.I., Nathanson, D., Marquis, P.M. (2002). Are adhesive technologies needed to support ceramics? An assessment of the current evidence. *J. Aesthet. Dent.* 4, 7-22.

Castro, H.L., Corazza, P.H., Paes-Júnior, T.A., Della Bona, A. (2012). Influence of Y- TZP ceramic treatment and different resin cements on bond strength to dentin. *Dent. Mater.* 28, 1191-7.

Cavalcanti, A.N., Foxton, R.M., Watson, T.F., Oliveira, M.T., Giannini, M., Marchi, G.M. (2009). Bond strength of resin cements to a zirconia ceramic with different surface treatments. *Oper. Dent.* 34, 280-287.

Cheung, G.J., Botelho, M.G. (2015). Zirconia Surface Treatments for Resin Bonding. *J. Adhes. Dent.* 17, 551-558.

Cura, C., Özcan, M., Isik, G., Sracoglu, A. (2012). Comparison of alternative adhesive cementation concepts for zirconia ceramic: glaze layer vs zirconia primer. *J. Adhes. Dent.* 14, 75-82.

Curtis, A.R., Wright, A. J., Fleming, G.J.P. (2006). The influence of surface modification techniques on the performance of a Y-TZP dental ceramic. *J. Dent.* 3, 195-206.

Della Bona, T.A., Donassollo, F.F., Demarco, A.A., Barrett, A.A., Mecholsky Jr, J.J. (2007). Characterization and surface treatment on topography of a glass-infiltrated alumina/zirconia- reinforced ceramic. *Dent. Mater.* 23, 769-775.

Derand, T., Molin, M., Kleven, E., Haag, P., Karlsson, S. (2008). Bond strength of luting materials to ceramic crowns and different surface treatments. *Eur. J. Prosthod. Restor. Dent.* 16, 35-38.

Derand, T., Molin, M., Kvam, K. (2005) Bond strength of composite luting cement to zirconia ceramic surfaces. *Dent. Mater.* 21, 1158-1162.

Derand. P., Derand, T. (2000). Bond strength of luting cements to zirconium oxide ceramics. *Int. J. Prosthod.* 13, 131-5.

Duran, P., Moure, C. (1984). Sintering at near theoretical density and properties of PZT ceramics chemically prepared. *J. Mater. Sci.* 20, 827-833.

Ehlers, V., Kampf, G., Stender, E., Willershausen, B., Ernst, C.P. (2015). Effect of thermocycling with or without 1 year of water storage on retentive strengths of luting cements for zirconia crowns. *J. Prosthet. Dent.*113, 609-615.

El-Korashy, D.I., El-Refai, D.A. (2014). Mechanical properties and bonding potential of partially stabilized zirconia treated with different chemomechanical treatments. *J. Adhes. Dent.* 16, 365-76.

Erdem, A., Akar, G.C., Erdem, A., Kose, T. (2014). Effects of different surface treatments on bond strength between resin cements and zirconia ceramics. *Oper. Dent.* 39, e118-e127.

Ferrari, M., Vichi, A., Zarone, F.(2015). Zirconia abutments and restorations: from laboratory to clinical investigations. *Dent. Mater.* 31, e63-76.

Garvie, R.C., Nicholson, P.S. (1972). Structure and thermomechanical properties of partially stabilized zirconia in the CaO-ZrO_2 system. *J. Amer. Ceram. Soc.* 55, 152-157.

Guan, S.H., Zhang, X.J., Liu, Z.P. (2015). Energy landscape of zirconia phase transitions. *J. Am. Chem. Soc.*, 137, 8010-8013.

Guazzato, M., Quach, L., Albakry, M., Swain, M.V. (2005). Influence of surface and heat treatments on the flexural strength of Y-TZP dental ceramic. *J. Dent.* 33, 9-18.

International Standards Organization: Implants for surgery. Ceramic materials based on yttria-stabilized tetragonal zirconia. 2008.

Janda, R. Roulet, J.F., Wulf, M., Tiller, H.J. (2003). A new adhesive technology for all-ceramics. *Dent. Mater.* 19, 567-573.

Karakoca, S., Yilmaz, H. (2009). Influence of surface treatments on surface roughness, phase transformation, and biaxial flexural strength of Y-TZP ceramics. *J. Biomed. Mater. Res. Part B: App. Biomater.* 91B, 930-937.

Kasraei, S., Rezaei-Soufi, L., Heidari, B., Vafaee, F. (2014). Bond strength of resin cement to CO_2 and Er:YAG laser-treated zirconia ceramics. *Restor. Dent. Endod.* 39, 296-302.

Kern, M., Barloi, A., Yang, B. (2009). Surface conditioning influences zirconia ceramic bonding. *J. Dent. Res.* 88, 817-822.

Kern, M., Wegner, S.M. (1998). Bonding to zirconia ceramic: adhesion methods and their durability. *Dent. Mater.* 14, 64-71.

Kim, S.T., Cho, H.J., Lee, Y.K., Choi, S.H., Moon, H.S. (2010). Bond strength of Y-TZP-zirconia ceramics subjected to various surface roughening methods and layering porcelain. *Surf. Inter. Analysis* 42, 576-80.

Kosmac, T., Oblak, C., Jevnikar, P., Funduk, N., Marion L. (2000). Strength and reliability of surface treated Y-TZP dental ceramics. *J. Biomed. Mater. Res.* 53, 304-313.

Kosmac, T., Oblak, C., Jevnikar, P., Funduk, N., Marion, L. (1999). The effect of surface grinding and sandblasting on flexural strength and reliability of Y-TZP zirconia ceramic. *Dent. Mater.* 15, 426-433.

Liu, D., Pow, E.H., Tsoi, J.K.H., Matinlinna, J.P. (2014). Evaluation of four surface coating treatments for resin to zirconia bonding. *J. Mech. Behav. Biomed. Mater.*, 32, 300-309.

Liu, D., Tsoi, J.K.H., Matinlinna, J.P., Wong, H.M. (2015). Effects of some chemical surface modifications on resin zirconia adhesion. *J. Mech. Behav. Biomed. Mater.*, 46, 23-30.

Lughi, V., Sergo, V. (2010). Low temperature degradation -aging- of zirconia: A critical review of the relevant aspects in dentistry. *Dent. Mater.* 26, 807-820.

Lung, C.Y.K., Matinlinna, J.P. (2012). Aspects of silane coupling agents and surface conditioning in dentistry: an overview. *Dent. Mater.* 28, 467-477.

Matinlinna, J.P., Heikkinen, T., Özcan, M., Lassila, L.V., Vallittu, P.K. (2006). Evaluation of resin adhesion to zirconia ceramic using some organosilanes. *Dent. Mater.* 22, 824-31.

Mattiello, R.D.L., Coelho, T.M.K., Insaurralde, E., Coelho, A.A. K., Terra, G.P., Kasuya, B., Favar I.N, Gonçalvez, L.S., Fonseca, R. B. (2013). A review of surface treatment methods to improve the adhesive cementation of zirconia-based ceramics. *ISRN Biomaterials*, 2013.

Michida, S.M., Kimpara, E.T., dos Santos, C., Souza, R.O., Bottino, M.A., Özcan, M. (2015). Effect of air-abrasion regimens and fine diamond bur grinding on flexural strength, Weibull modulus and phase transformation of zirconium dioxide. *J. Appl. Biomater. Func. Mater.* 16, e266-273.

Miyazaki, T., Nakamura, T., Matsumura, H., Ban, S., Kabayashi, T. (2013). Current status of zirconia restoration. *J. Prosthodont. Res.* 57, 236-261.

Monticelli, F., Osorio, R., Mazzitelli, C., Ferrari, M., Toledano, M. (2008). Limited decalcification / diffusion of self-adhesive cements into dentin. *J. Dent. Res.*87, 974-979.

Oyagüe, R.C., Monticelli, F., Toledano, M., Osorio, E., Ferrari, M., Osorio, R. (2009). Influence of surface treatments and resin cement selection on bonding to densely sintered zirconium-oxide ceramic. *Dent. Mater.* 25, 172-179.

Özcan, M. (2013). Air abrasion of zirconia resin-bonded fixed dental prostheses prior to adhesive cementation: why and how. *J. Adhes. Dent, 15,* 394.

Özcan, M. (2014). Airborne particle abrasion of zirconia fixed dental prostheses. *J. Esthet. Restor. Dent.* 26, 359-362.

Özcan, M., Bernasconi, M. (2015). Adhesion to zirconia used for dental restorations: a systematic review and meta-analysis. *J. Adhes. Dent.* 17, 7–26.

Özcan, M., Melo, R.M., Souza, R.O., Machado J.P., Valandro L., Botttino M.A (2013). Effect of air-particle abrasion protocols on the biaxial flexural strength, surface characteristics and phase transformation of zirconia after cyclic loading. *J. Mech. Behav. Biomed. Mater.* 20, 19-28.

Özcan, M., Nijhuis, H., Valandro, L.F. (2008). Effect of various surface conditioning methods on the adhesion of dual-cure resin cement with MDP functional monomer to zirconia after thermal aging. *Dent. Mater. J.* 27, 99-104.

Özcan, M., Vallittu, P.K. (2003). Effect of surface conditioning methods on the bond strength of luting cements to ceramics. *Dent. Mater.* 19:825-831.

Özcan, M.; Volpato, C.A.M. (2015). Adhesion to zirconium dioxide used for dental reconstructions: surface conditioning concepts, challenges and future prospects. *Cur. Oral Health Rep.* 2, 190-194.

Passos, S.P., May, L.G., Barca, D.C., Özcan, M., Bottino, M.A., Valandro, L.F. (2010). Adhesive quality of self-adhesive and conventional adhesive resin cement to Y-TZP ceramic before and after aging conditions. *Oper. Dent.* 35, 689-696.

Passos, S.P., Nychka, J.A., Major, P., Linke, B., Flores-Mir, C. (2015). In vitro fracture toughness of commercial Y-TZP ceramics: A systematic review. *J Prosthodont.* 24, 1-11.

Piascik, J.R., Swi, E.J., Thompson, J.Y., Grego, S., Stoner, B.R. (2009). Surface modification for enhanced silanation of zirconia ceramics. *Dent. Mater.* 25, 1116-1121.

Piconi, C., Maccauro, G. (1999). Zirconia as a ceramic biomaterial. *Biomaterials* 20, 1-25.

Reich, S., Hornberger, H. (2002). The effect of multicolored machinable ceramics on the esthetics of all-ceramic crowns. *J. Prosthet. Dent.* 88, 44-49.

Sarmento, H.R., Campos, F., Souza, R.S., Machado, J.P., Souza, R.O., Bottino, M.A., Özcan, M. (2014). Influence of air-particle deposition on the surface topography and adhesion of resin cement to zirconia. *Acta Odontolol. Scand.* 72, 346-353.

Sciasci, P., Abi-Rached, F.O., Adabo, G.L., Baldissara, P., Fonseca, R.G. (2015). Effect of surface treatments on the shear bond strength of luting cements to Y-TZP ceramic. *J. Prosthet. Dent.* 113, 212-219.

Senyilmaz, D.P., Palin, W.M., Shortall, A.C., Burke, F.J. (2007) The effect of surface preparation and luting agent on bond strength to a zirconium-based ceramic. *Oper. Dent.* 32, 623-630.

Souza, R.O., Valandro, F.L., Melo, R.M., Machado, J.P., Bottino,M.A., Özcan, M.(2013). Air-particle abrasion on zirconia ceramic using different protocols: effects on biaxial flexural strength after cyclic loading, phase transformation and surface topography. J. *Mech. Behav. Biomed. Mater.* 26, 155-63.

Subaşı, M.G., Özgür, I. (2014). Influence of surface treatments and resin cement selection on bonding to zirconia. *Lasers Med. Sci.* 29, 19-27.

Thompson, J.Y., Stoner, B.R., Piascik, J.R., Smith, R. (2011). Adhesion/ cementation to zirconia and other non-silicate ceramics: where are we now? *Dent. Mater.* 27, 71-82.

Tsuo, Y., Yoshida, K., Atsuta, M. (2006). Effects of alumina-blasting and adhesive primers on bonding between resin luting agent and zirconia ceramics. *Dent. Mater. J.* 25, 669-674.

Turp, V., Sen, D., Tuncelli, B., Goller, G., Özcan, M. (2013). Evaluation of air-particle abrasion of Y-TZP with different particles using microstructural analysis. *Austr. Dent. J.* 58, 183-191.

Tzanakakis, E.G.C., Tzoutzas, I.G., Koidis, P.T. (2016). Is there a potential for durable adhesion to zirconia restorations? A systematic review. *J. Prosthet. Dent.*, 115, 9-19.

Vagkopoulou, T. Koutayas, S.O., Koidis, P., Strub, J.R. (2009). Zirconia in dentistry: Part 1. Discovering the nature of an upcoming bioceramic. *Eur. J. Esthet. Dent.* 4, 130-151.

Valandro, L.F., Özcan, M., Amaral, R., Leite, F.P., Bottino, M.A. (2007). Microtensile bond strength of a resin cement to silica-coated and silanized In-Ceram zirconia before and after aging. *Int. J. Prosthodont.* 20, 70-72.

Vrochari, A.D., Eliades, G., Hellwig, E., Wrbas, K.T. (2009). Curing efficiency of four self-etching, self-adhesive resin cements. *Dent. Mater.* 25, 1104-1108.

Xible, A.A., Jesus Tavarez, R.R., Araujo, C.R., Bonachela, W.C. (2006). Effect of silica coating and silanization on flexural and composite-resin bond strengths of zirconia posts: an in vitro study. *J. Prosthet. Dent.* 95, 224-229.

Zhang, Y., Kim, J.W. (2009). Graded structures for damage resistant and aesthetic all-ceramic restorations. *Dent. Mater.* 25, 781-790.

Zhang, Y., Lawn, B.R., Rekow, E.D., Thompson, V.P. (2004). Effect of sandblasting on the long-term performance of dental ceramics. *J. Biomed. Mater. Res. Part B: Appl. Biomater.* 15, 381-386.

Zhang,Y., Lawn, B.R., Malament, K.A., Thompson, V.P., Rekow, E.D. (2006). Damage accumulation and fatigue life of particle-abraded ceramics. *Int. J. Prosthodont.* 19, 442-448.

INDEX

D

E

F

G

H

I

L

M

N

O

P

Q

R

S

T

U

V

W

X

Y

Z